COLORADO
LAKES & RESERVOIRS
FISHING & BOATING GUIDE

OUTDOOR BOOKS & MAPS

Outdoor Books & Maps
An imprint of
APC Enterprises, LLC
PO Box 2948
Parker, CO 80134

(303) 660-2158
(800) 660-5107
FAX: (303) 688-4388

Contents

There's a special place waiting for you, where the air tastes sweeter and excitement is plentiful. The price of admission is cheap, yet the rewards are priceless. Welcome to the rivers, lakes and reservoirs of Colorado, home to some of the finest fishing in the world.

Fishing in Colorado means an almost limitless amount of choices. It can put you among towering snow-capped peaks or against the backdrop of stately high-rises. If you want the convenience of easy access urban fisheries, you've no better place than Colorado's Front Range. Sprinkled throughout the greater metro areas are dozens upon dozens of well-stocked, highly productive habitats that have produced more than one record catch.

Or perhaps your fancy is a cool, secluded alpine stream where brook trout feed in crystal clear pools of mountain run-off. Again, your options are virtually infinite and governed only by your imagination. Whether you're a deep water troller, an ardent fly fisherman or simply a weekend fisherman who frequents close-to-home waters, Colorado has it all.

Your choices aren't limited to location: there are 38 recorded species of fish in the state! Yet mention Colorado in the same breath as fishing and the likely response evoked will be "trout." And for good reason. The lower 48's most mountainous state provides high-elevation rivers, lakes, streams and reservoirs that are ideal waters for these cold water fish. Rainbow, cutthroat, brook, lake, brown and golden trout all occupy various cold water habitats throughout Colorado. Trout, and increasingly kokanee salmon, have come to collectively symbolize Colorado's sportfishing species of choice.

However, through the stocking programs administered by the Colorado Division of Wildlife, an abundance of warm water fish also call Colorado their home. Midwest favorites like largemouth and smallmouth bass, walleye, yellow perch, bluegill, crappie, catfish, along with the hybridized wiper and tiger muskie, provide anglers an unusually diverse range of fishing opportunities.

The beauty of fishing in Colorado is more than meets the eye. Not only is the scenery par excellence, but pristine mountain lakes and expansive flatland reservoirs are little more than an hour's drive apart. The same holds true for the myriad urban lakes and ponds, and the outlying reaches of larger interstate waterways like the South Platte and Arkansas rivers. Few places offer the near proximity of warm water and cold water fishing as does Colorado.

To fully enjoy fishing in Colorado is to sample its array of offerings. Almost any location is sure to provide some degree of fishing action, but as anyone who's ever wetted a line knows, locations vary in terms of fishing potential. The best advice to connect you with your fish will come via a visit to a bait and tackle shop, preferably one that's close to the area you'll be fishing or that has a reputation which exceeds the shop's backyard roost. Read the weekly fishing reports for the scoop on the bite, pay a few dollars for the current lures and baits of choice, and get the Colorado Division of Wildlife's Fishing Information Directory — issued when you apply for your fishing license. The DOW directory is updated every couple years and tells you of your responsibilities, the requirements and regulations concerning the catching of fish in Colorado.

Preparation and planning will make your Colorado fishing experience a success. Because of the diversity of the state's fisheries, it's important to know what you'll be fishing for ahead of time. While the occasional anglers may fare quite well simply by fishing the basic worm and bobber combination or indiscriminately tossing any lure, it's a near guarantee the results will be tripled if familiarization with the species being sought is part of the game plan. Take the time to acquaint yourself with the characteristics and habitats of your game fish, whether you're fishing self-sustaining or regularly stocked waters, etc. With a more targeted approach, you'll probably benefit with a higher catch ratio and derive greater pleasure knowing your tactics worked.

Index Map

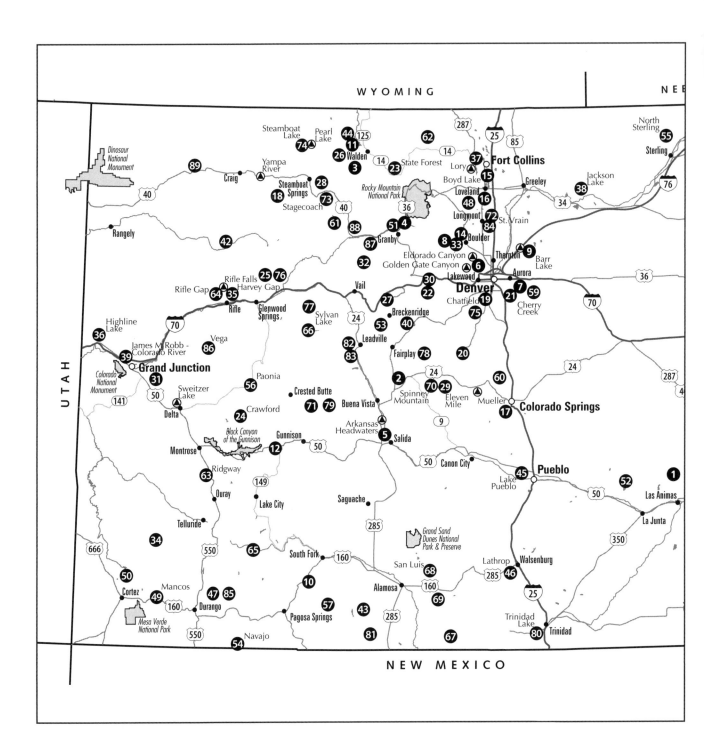

Index Map

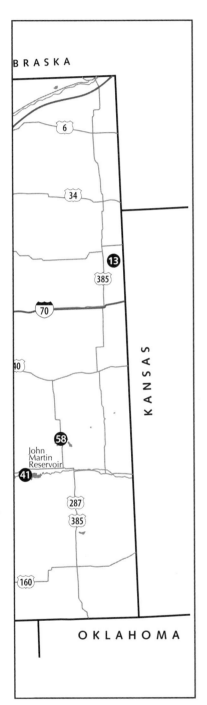

Fish You are Likely to Catch

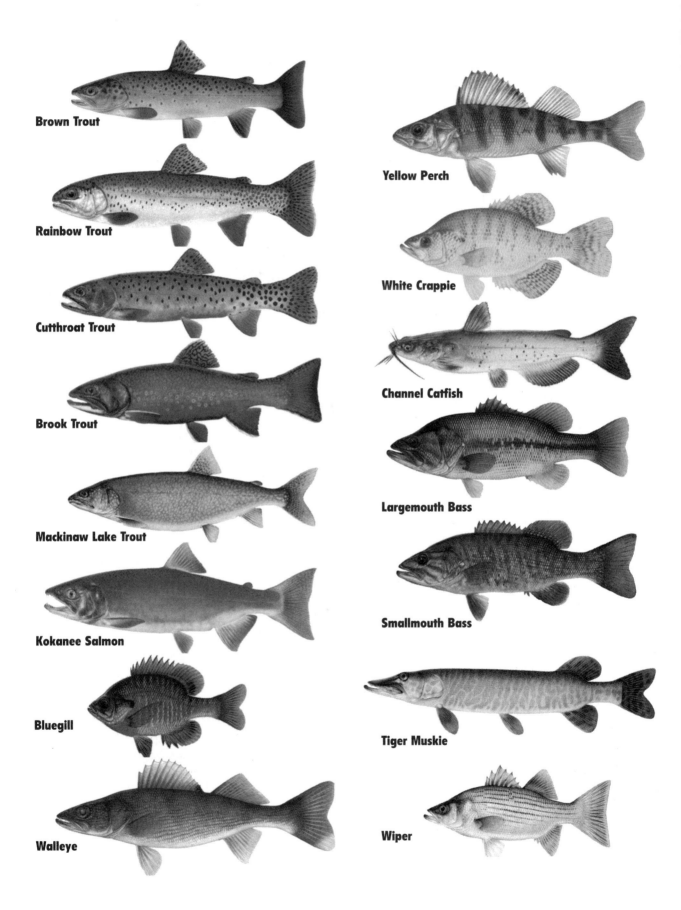

Brown Trout

Rainbow Trout

Cutthroat Trout

Brook Trout

Mackinaw Lake Trout

Kokanee Salmon

Bluegill

Walleye

Yellow Perch

White Crappie

Channel Catfish

Largemouth Bass

Smallmouth Bass

Tiger Muskie

Wiper

If you want to catch fish in Colorado, you must first become familiar with the various species that the state has to offer. There are nine cold water fish, most of which are trout, and six warm water fish that you should get to know.

Trout are more difficult to catch than many of their warm water cousins. Trout rely on sight to feed rather than smell techniques that are typical of many warm water fish such as catfish and panfish.

Trolling and casting are common methods to attract fish. Stream and river fishing involves the use of artificial flies and lures that can be carried by the current to fish in as natural a process as possible. Trolling is a common technique used on lakes.

Fish are one of the lowest forms of animal life. They have a relatively small brain for their body size. However, what they may be missing in intelligence, is more than made up for in keen eyesight.

Most fresh water fish feed at three basic times during the day. Morning feeding starts at sunrise and continues until 9 or 10 a.m. Feeding drops off until about noon when a short feeding frenzy will last until about 2 p.m. Late afternoons are the slowest time to fish. Feeding will start up again around sunset and can continue into the late evening.

Brown Trout

Brown trout were imported from Europe into the United States in 1883. They were planted in the streams and tributaries of the Colorado Rockies and have survived ever since.

They are one of the strongest survivors of the trout specie. They can tolerate warm, muddy water that would kill most other trout. Browns prefer slow moving cold water that has lots of large deep holes. Brown trout do most of their feeding at night. They will feed early in the morning but will move to the deep water and pools at sunrise.

Most of their feeding is done on the bottom. Browns are one of the more aggressive trout and will eat just about any aquatic form of life, small swimming birds and mammals.

Spawning season for browns is in the fall. They will congregate in headwater streams. The female will produce from 200 to 600 eggs in 10 to 30 feet of water.

Rainbow Trout

The rainbow trout is one of the scrappiest fish in the trout family, and hence a favorite with fishermen. Rainbow trout were originally a native of the West Coast. The trout was introduced into Colorado in the 1880s and has done extremely well ever since. In fact, about 75% of the trout caught in the state are rainbows.

Rainbows are notorious for being aggressive strikers. They are attracted to artificial flies and bright colored lures. They like lower, warmer waters. Rainbows spawn in the spring.

Brook Trout

Many consider the brook trout to be the prettiest of all the trout. Their body is distinguished by side markings which are spattered with red and white spots on a dark green background.

Cutthroat Trout

Cutthroat trout are the only trout know to be native to Colorado. Their name is derived by the fact that they have a splash of red behind the gills.

Cutts, as they are fondly called, are found in the upper stretches of the mountain streams and lakes. They do not adapt well to warm, silty water. If you want to catch one of these fish, you'll have to be prepared to hike into their habitat.

These fish will go after any of the artificial materials including flies and lures. If you hook one, chances are it is the first time that they have ever seen what your are using. Cutts move into the headwaters of mountain streams in the spring to spawn.

Mackinaw Lake Trout

The Mackinaw can be found in many of Colorado's deep, cold water lake. It is by far the largest trout in Colorado's trout inventory. Mackinaw in excess of 60 pounds have been consistently caught.

Mackinaw are the most difficult trout to catch because they spend most of the year in deep water. When we say deep water, we are talking about depths in excess of 60 feet. They move to the shallow water to feed in the early spring, right after "ice-off' and in the fall to spawn, which is the best time to catch these fish.

Most fish are caught on deep rigged trolling gear. They will also hit salmon eggs that are cast from shore or dropped from a boat. Winter ice fishermen will attach sucker meat to Airplane jigs dropped to 60 plus feet to successfully catch Mackinaw.

Whole dead suckers also work just as well during the spring and summer. Weight the sucker to help get it down near the bottom where you want it.

How to Catch Colorado Fish

The next step is to rig the fish so that when you're done, it will be rigged like an artificial plug. This is accomplished by attaching two No. 2 or 4 treble hooks to the front and posterior of the fish. The hooks can be either tied onto the sucker or you can use a needle to thread the line through the sucker and attach the hooks to the end of the line.

Grayling

Grayling are probably the least known trout in the Colorado trout family. When fishermen mention grayling, most think of Canada and Alaska. They can be found only in selected regions of high Colorado mountain lakes and cool streams.

These fish are very fond of insects and most often are caught by fly fishermen. They can also be attracted by the typical small trout lures such as the Kasmasters. Grayling will mature at about 20 inches or 4 pounds. They like to swim in schools, which can add to your fishing pleasure if you find the school. Spawning begins in March and ends in June.

Kokanee Salmon

Kokanee are land-locked pacific sockeye salmon. They were introduced into Colorado in the early fifties and inhabit many of the same lakes that are popular with trout fishermen. Kokanee will often be taken while fishing for trout. They will go after all of the popular trout lures and are equally fond of their own salmon eggs and other common trout baits.

These fish like to surface feed during the day. They will move down to deep water at night and move back up to the surface at sunrise. Most kokanee are taken when trolling. Spawning begins in October and runs through December. The fish school up in the tributaries that flow out of their home lakes. Colorado allows fishermen to snag kokanee with weighted treble hooks during the spawning run.

Mountain White Fish

Whitefish live in Colorado's faster and larger rivers. They are particularly prevalent in the Yampa and White Rivers where they can be found in the large pools. These fish feed at night and prefer insect and larvae. Spawning begins in the fall.

Bluegill

Bluegills were introduced to Colorado in the early 1920s. They abound in many of the state's lower lakes and streams and are considered a primary food source for trout and bass. They offer fishermen fast-paced fishing action and are excellent eating.

They are particularly active in the early morning hours and late evening. Favorite baits are worms and small dry flies. Bluegill like to feed in shallow water. However, as the water warms up in the later summer season, they will move down to deeper water.

Spawning season begins in the spring and continues until late August. Nests are built in colonies by the male of the species who also guards over their spouse's nest.

Walleye

Walleye are the largest member of the perch family and can be found in many of Colorado's larger reservoirs. A typical walleye will run anywhere from 2 to 10 pounds.

The name walleye was derived from the large bulging eyes that are a dominant characteristic of this fish. Their eyes are very sensitive to light. As a result, they will move down into deep water as soon as the morning sun hits the water surface.

Walleye feed at night. They will prowl the shoreline in search of minnows, their favorite food. Casting in the dark with artificial or live minnows is the best way to catch these fish. Spawning begins in the early spring.

Perch

Perch are relatively small fish that rarely exceed a pound. They are found in many of Colorado's warm water reservoirs and live in schools. They will spend most of the day in deep water. As evening approaches, they will move into shallow water to feed.

The best time to catch perch is from noon on into the evening. Fish with small flies, lures or natural baits a foot or two off the bottom for best results. Perch spawn in the spring.

Crappie

Crappies were introduced into Colorado waters in the 1880s and prefer the warmer reservoirs. They like to gather in schools around submerged brush and rock areas. During the summer months they will stay in about 8 to 15 feet of water.

Live minnows with a bobber are one of the most popular ways to fish for crappie. Fishing for crappie requires sensitive tackle to be successful. A small spin cast bubble can work better than a standard bobber. The line can be slipped through the center of the bubble, and you can add water to the inside of the bubble to improve on casting distances. A light-hitting crappie will be less likely to drop the bait if it doesn't feel the initial pull of the bubble.

Catfish

Catfish are native to Colorado and inhabit many of the state's warmer reservoirs and rivers. These fish like to feed at night. Using natural baits such as night crawlers, crayfish, chicken livers or scented dough balls catches most catfish.

Catfish can grow in excess of 50 pounds. Most will average between 2 and 5 pounds. Catfish have a keen sense of smell, which is the reason why they will go after odorized bait.

Catfish move out of the reservoirs and into the spillway canals during Colorado's irrigation season, which starts in the late spring. They like the moving water and find plenty of feed in the runoff.

Look for the deeper pools to fish. Cast minnows with enough weight to carry them to the bottom. If the cats are there, you'll hook-up within 15 minutes. If you get no immediate action, move on to the next pool.

Bass

Bass are a member of the sunfish family and enjoy the reputation of being furious fighters when hooked. The two species of bass that are prevalent in Colorado are the large and smallmouth bass. Largemouth were introduced into the state in the late 1800s. Smallmouth were introduced in the early 1950s.

Largemouth are larger than their smallmouth cousin. Adults will weigh up to 10 pounds or more. A typical smallmouth will average about 2 pounds.

Bass prefer shallow weed infested lakes. Bass action begins to heat up in late spring, just before the spawn. Bass become very aggressive and hungry during spawning. They can be found cruising the shoreline for food in the early morning and late evening hours. Casting and spin fishing are the best ways to catch a bass.

An important point to remember about bass is that they will adjust to the sun reflections on the water by changing their swim depth. The higher the sun's intensity, the deeper the depth that you will find bass.

Start shallow and work incrementally into deeper water. You are bound to find bass sooner or later. Once you get your first hit, mark the depth that you were at. The rest of the bass will be at that same level.

Northern Pike

Next to Minnesota, Colorado is one of the better pike fishing states in the nation. Pike were imported into the state in the mid-1950s to help control a rising sucker population in many lakes. Unfortunately, the pike developed an appetite for trout and left the suckers alone.

A mature pike will weigh anywhere from 3 to 15 pounds. They like shallow water with abundant vegetation. However, they will move into deeper water during extremely warm or cold weather.

The best time to catch a pike is in late spring, right after ice off. Most pike are caught casting big spoons and plugs into the shallow water near weed beds. Attach 6 to 8 inches of steel leader to 10 to 20 pound test line.

If possible, wade into the weeds and cast at different angles out into the open water. Retrieve the plug or spoon just over the top of the weeds. If you get a missed strike, stop reeling for a few seconds. Pike have a tendency to want to continue an attack a second time.

-David Rye

Section 1

Major Colorado Lakes and Reservoirs

Adobe Creek Reservoir

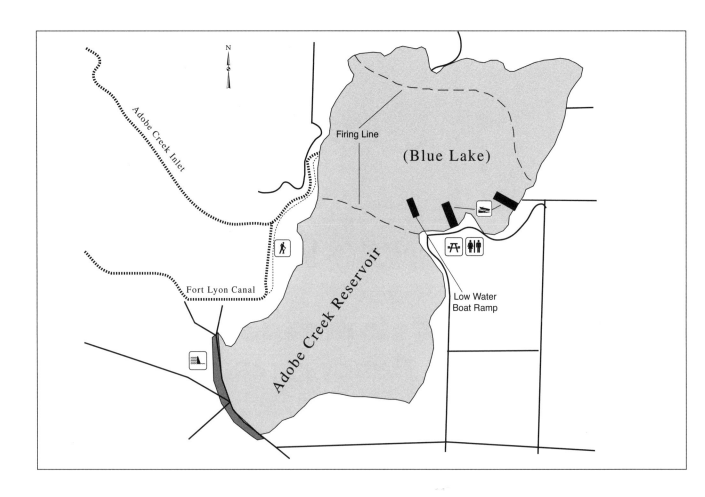

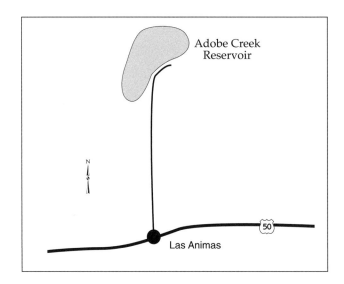

DIRECTIONS From Denver, take I-25 south of Pueblo, then travel east on US 50 to Las Animas. Travel north 12 miles on County Road 10 to the reservoir.

FEE No.

SIZE 5,029 acres when full.

ELEVATION 4,128 feet.

MAXIMUM DEPTH 35 feet.

FACILITIES Restrooms, boat ramps.

BOAT RAMP Two ramps are available depending on the variable water level. The quality of the ramps, which are mostly concrete, also depends on the water level.

FISH Largemouth bass, bluegill, walleye, wiper, crappie, and tiger muskie. Crappie; trolling lead-headed jigs in white or chartreuse with part of a nightcrawler, twister tail or jigs and minnows near any brush lines. Bass; top-water spinner baits, crank baits and Texas-rigged nightcrawlers or plastic worms. Walleye and saugeye; trolling bottom bouncers with nightcrawlers or minnows, deep-running Rapala or Rebel lures resembling minnows.

RECREATION Boating, fishing, sailing, personal watercraft, swimming, waterskiing.

CAMPING Open camping southeast side of reservoir. There is no shade and no trees, and the area is often windy.

CONTROLLING AGENCY Division of Wildlife.

INFORMATION Lamar Office (719) 336-6600.

MAP REFERENCES Benchmark's *Colorado Road & Recreation Atlas:* p. 104.

Antero Reservoir

DIRECTIONS From Hartsel, drive 5 miles southwest on US 24.

FEE No.

SIZE 2,500 acres.

ELEVATION 8,940 feet.

FACILITIES Restroom, picnic tables, campground.

BOAT RAMP Two concrete ramps located on south and north sides. Boating is prohibited from 9 p.m.-4 a.m.

FISH Rainbow, brown and cutthroat trout. The shoreline is shallow so waders are essential. It can be a good spin-caster's lake when the wind isn't blowing. Good results can be achieved by bait fishing from shore with a long cast. Antero produces the best fishing just after spring thaw, and the reservoir will have heavy aquatic growth during the summer. Best results for shore fishing will be achieved if bait is presented just off the bottom. Ice fishing shelters must be portable.

RECREATION Fishing, boating, sailing, ice fishing, hunting and wildlife viewing.

CAMPING Open camping is available in designated camp areas. Antero is part of the South Park Region. The area is devoid of trees and shrubs and is known for being windswept.

CONTROLLING AGENCY Division of Wildlife.

INFORMATION Colorado Springs Office (719) 227-5200.

MAP REFERENCES Pike National Forest map. Benchmark's *Colorado Road & Recreation Atlas:* p. 87.

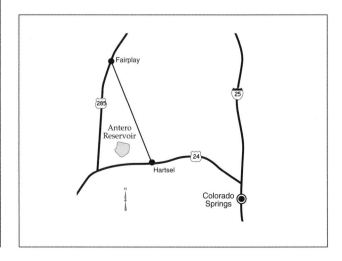

Arapaho Lakes

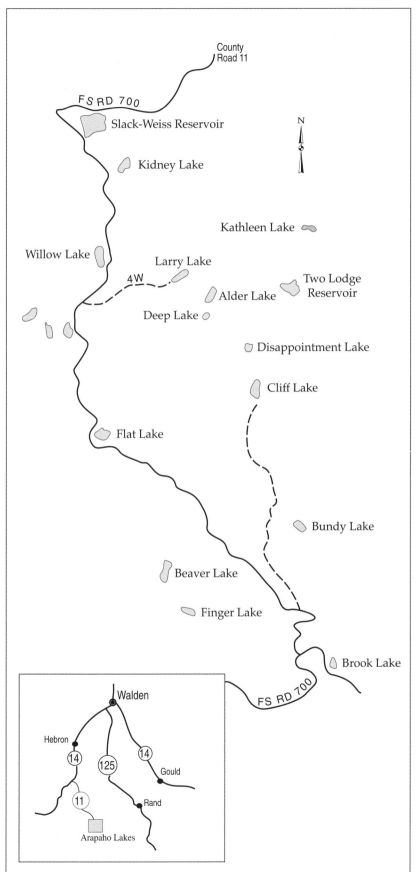

DIRECTIONS Take I-25 north to Fort Collins, then travel west on Colorado 14, over Cameron Pass through the town of Walden. Access to the area is south of Walden on Colorado 14, toward the town of Hebron. About 4 miles south of Hebron, travel east on County Road 28 for one mile and then south on County Road 11, which takes you into Medicine Bow & Routt National Forests.

FEE No.

SIZE This area is made up of several small lakes and ponds, including Bundy Lake (3 acres), Willow Lake (4 acres), Kathleen Lake (2.5 acres), Kidney Lake (3 acres), Slack Weiss Reservoir (14 acres), Two Ledge Reservoir (6 acres), Flat Lake (5.5 acres) and Disappointment Lake (1.5 acres). Other lakes in the vicinity are Deep, Finger, Beaver, Long, Alder, Round and Cliff.

ELEVATION Ranges from 9,000 feet to 10,000 feet.

FACILITIES Outhouse at Slack Weiss.

BOAT RAMP No ramps are available, hand-launch only. The US Forest Service does not have any restrictions as to the types of boats used. Slack Weiss Reservoir is restricted to electric motors.

FISH All of the lakes are good for natural species such as brook, brown and cutthroat trout. A few of the lakes also have stocked catchable rainbow. Typical bait are lures or flies, depending on the hatch.

RECREATION Fishing, boating.

CAMPING The area is open "Leave No Trace" camping. Camping is not allowed within 100 feet of any lake or stream. Refer to Routt National Forest Service map for other campground information.

CONTROLLING AGENCY Medicine Bow & Routt National Forests.

INFORMATION Parks Ranger District (970) 723-8204.

MAP REFERENCES Routt National Forest map.
Benchmark's *Colorado Road & Recreation Atlas:* p. 60.

Arapaho National Recreation Area

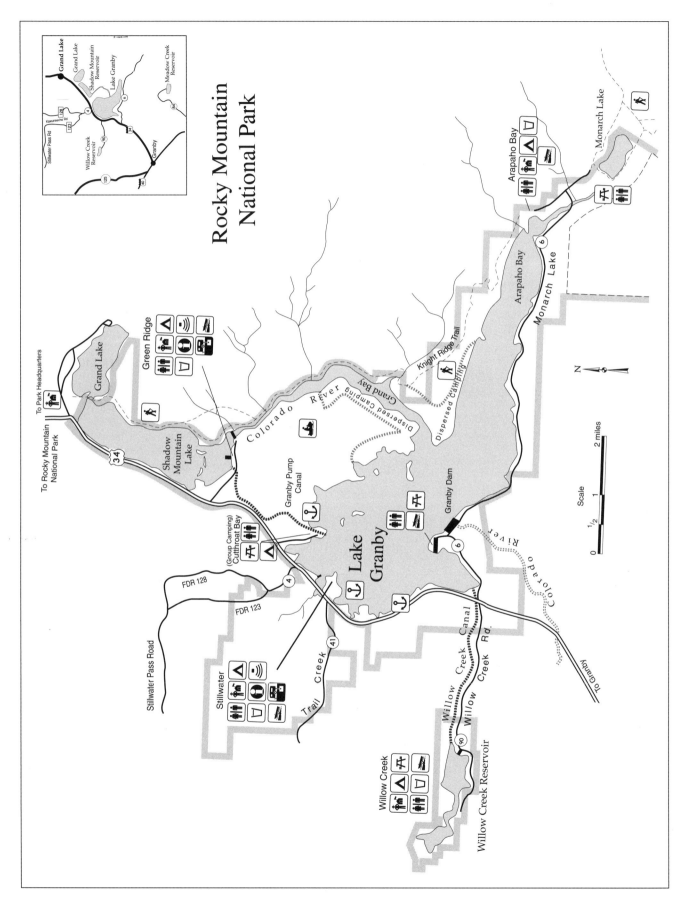

DIRECTIONS From Denver, take I-70 west to exit 232 (US 40). Head north on US 40 to the town of Granby. On the west side of Granby, turn north at junction for US 34 north.

FEE Daily or annual pass.

SIZE Grand Lake - 506 acres; Lake Granby - 7,256 acres; Monarch Lake - 147 acres; Shadow Mountain - 1,400 acres; Willow Creek Reservoir - 750 acres; Meadow Creek Reservoir - 50 acres.

ELEVATION 8,367 feet.

MAXIMUM DEPTH Grand Lake - 265 feet; Lake Granby - 221 feet; Monarch Lake - 30 feet; Shadow Mountain - 24 feet.

FACILITIES Dump stations at Stillwater and Green Ridge Campgrounds, toilets, picnic tables and fire grates. Beginning Memorial Day weekend and through the summer, running water also is available.

BOAT RAMP Boat ramps are at the Stillwater Campground, Sunset Point, Arapaho Bay Campground, Green Ridge Campground, Willow Creek Campground, and Hilltop picnic area. Some ramps are concrete. Marinas with fee ramps are located around some of the lakes.

RECREATION Fishing, boating, sailing, hiking.

CAMPING 325 campsites in four campgrounds and two group campsites are available. Stillwater Campground has 129 sites, Arapaho Bay Campground 84 sites, Greenridge 78 sites, and Willow Creek 34 sites. Unreserved sites are available on a first-come, first-serve basis. Reservations can be made for Stillwater, Arapaho Bay and Green Ridge Campgrounds.

FISH Lake Granby; Rainbow, brown and lake trout and kokanee salmon. Rainbow and brown trout fishing is best from shore using typical baits on the bottom. Lures also produce good action trolled or fished from shore. Lake trout are caught trolling or jigging at depths of 50-60 feet. As the water warms, the lake trout will tend to move to shallow water for a short time before moving back to deeper depths. Kokanee can

be caught trolling with single hook lures in fluorescent colors.

Willow Creek Reservoir; Rainbow and brook trout. This reservoir can produce excellent stocker rainbow with larger ones being taken occasionally. The best fishing is from shore in early mornings and late evenings on bait, rigged just off the bottom. Fair to good results are generally reported from slow trolling with flashers or Pop Gear rigged with half a nightcrawler. Wakeless speeds only. No restrictions on boat or motor size.

Shadow Mountain; Best for rainbow and brown through July, kokanee later in the summer. Can become weedy in the summer and tough to fish. Use typical baits and lures.

Grand Lake; Considered one of the best lake trout fisheries in the state with fish caught weighing 20 pounds or more. Rainbow, brown and kokanee are also present. Rainbow fishing is best in the spring. Wet flies and lures are good for brown and rainbow. Sucker meat for mackinaw.

The Arapaho National Recreation Area (NRA) is located in Grand County in the upper reaches of the Colorado River Valley. It was established by Congress on October 11, 1978, and is administered by the Forest Service—US Department of Agriculture. Most of the area was formerly part of Shadow Mountain NRA administered by the National Park Service.

The Arapaho NRA covers over 36,000 acres and contains the following major lakes: Lake Granby, Shadow Mountain Lake, Monarch Lake, Willow Creek Reservoir, and Meadow Creek Reservoir. Grand Lake, the largest natural lake in Colorado, is located outside the NRA, northeast of Shadow Mountain Lake.

The three largest lakes in the NRA (Shadow Mountain, Lake Granby and Willow Creek Reservoir) lie in the glacier sculpted valleys of the upper Colorado River. Most of the rolling terrain is composed of glacial debris and outwash. Many of the islands in the lakes are actually the tops of moraines, which are deposits of sand, gravel, and rock left by melting glaciers. Bedrock in most of the area is Precambrian crystalline rock or Tertiary volcanic rock.

Lodgepole pine and quaking aspen grow around the lakes, mixing with spruce and fir on mountain slopes. Wildlife species in the area include deer, moose, elk, coyote, bobcat, red fox, yellow bellied marmot, chipmunk, badger, pine squirrel, ground squirrel, beaver and black bear. Note: Black bear do visit campgrounds so make the necessary adjustments when visiting the area.

CONTROLLING AGENCY Arapaho & Roosevelt National Forests.

INFORMATION Sulphur Ranger District (970) 887-4100.

HANDICAPPED ACCESS Handicapped accessible facilities and fishing pier Point Park picnic area, Hilltop and Green Ridge Recreation Complex. Some facilities available with assistance.

MAP REFERENCES Arapaho & Roosevelt National Forests map.
Benchmark's *Colorado Road & Recreation Atlas:* p. 61.

Arkansas Headwaters Recreation Area

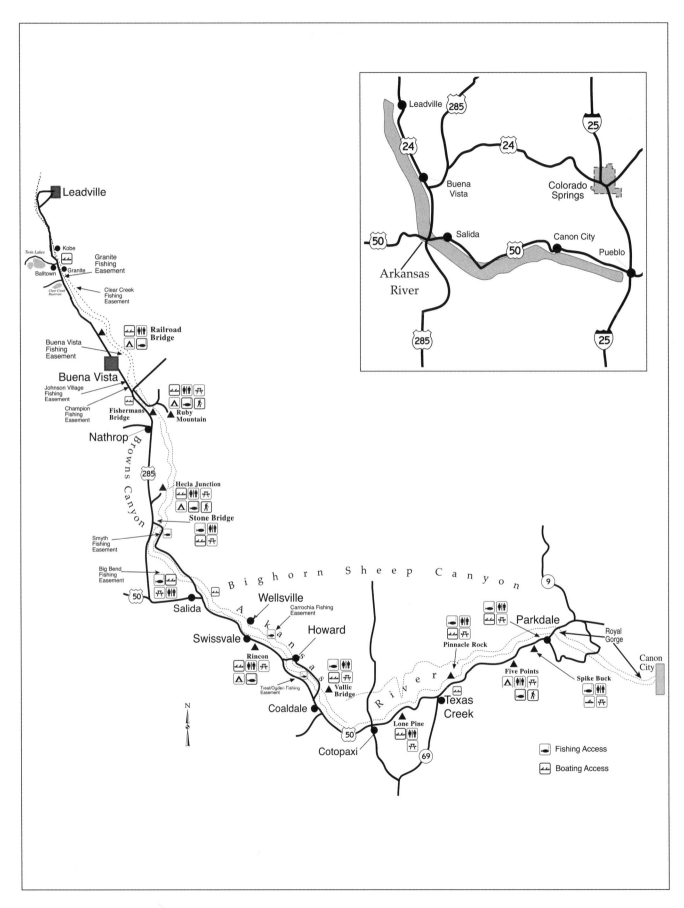

Fishing Access

Boating Access

DIRECTIONS Travel south from Denver on US 285 to the Colorado 291 cutoff just north of Salida. Travel south on Colorado 291 to US 50. At Salida, travel east along US 50 paralleling the river. Drive south from Denver on I-25 to Colorado Springs, then head southwest on Colorado 115 to Canon City. At Canon City, follow US 50 west along the river to Salida.

FEE Daily or Annual Pass.

ELEVATION Ranges from 4,900 feet at Pueblo Reservoir to 8,875 at Clear Creek Reservoir.

BOAT RAMP 14 boating access points from Nathrop to Parkdale.

FISH Trout, primarily rainbow and brown. The section of the river from Salida to Nathrop is known primarily for rainbow. While the best area for brown trout is from Salida to Parkdale, the area from the town of Howard to the stockyard bridge is known for larger rainbow trout. For bait fishermen, best results are generally found on night crawlers drifted along the bottom. Salmon eggs and dead minnows are always a good bet in the late evenings, as are live grasshoppers during the summer months. The best action for hardware fishermen is on Panther Martins in yellow/red/black or frog patterns as well as Rooster Tails, Kastmasters, Mepps, floating Rapala, Blue Fox and Hustler lures. During runoff, brighter colors produce best. Fishing access is clearly marked on the river. For fly fishermen, the best action is on Midge, Caddis and Royal Coachman patterns. Two sections, one just below Nathrop and between Texas Creek and Parkdale, are designated wild trout waters. Most of the river is open for bait as well as artificial lures. From Badger Creek (at the railroad trestle) to the stockyard bridge below Salida (a 7 1/2 mile stretch) and the Mugford Lease just north of Salida, artificial lures only. Check DOW regulations for bag limits and other restrictions.

RECREATION Fishing, rafting, hiking, sightseeing, gold panning, biking, picnicking.

CAMPING Five Points- 20 sites, Hecla Junction- 22 sites, Rincon- 8 sites, Railroad Bridge- 14 sites, Vallie Bridge-16 sites, and Ruby Mountain- 22 sites. Restrooms, picnic tables, grills, fishing access and parking.

HANDICAPPED ACCESS Handicapped accessible restrooms, visitors center, picnic tables, fishing, camping and trails. Lone Pine Recreation site offers a fishing trail. Some facilities available with assistance. Special needs fishing access at Blue Heron, Canon City, Hayden Meadows, Lone Pine, Stone Bridge, and Vallie Bridge.

OTHER INFORMATION
The upper Arkansas River Valley is a treasure of geology, history, wildlife, scenery and recreation opportunities. Run a rapid, explore a ghost town, hook a trout, watch a Bighorn sheep or view the Royal Gorge.

Buena Vista to Salida
This stretch's outstanding feature is Browns Canyon, a wilderness of pink granite punctuated with crashing whitewater rapids. This stretch offers fine fishing, as well as camping and picnicking. Ruby Canyon, the northern gateway, provides access to the 6,600 acre Browns Canyon Wilderness Study Area. Visitors can hike, backpack, camp, and view wildlife.

Below Browns Canyon, the valley widens again and the river calms down, though spectacular views remain. This stretch, called the Big Bend, offers prime trout fishing, and numerous Division of Wildlife easements give fishermen river access on private land. (Please observe posted rules when using fishing easements).

Salida to Vallie Bridge
Just below Salida the river flows into Bighorn Sheep Canyon, a granite canyon dotted with stands of pinon, juniper and oak brush. As the river flows into the Upper Arkansas Canyon, it veers slightly southeast, with the Sangre de Cristos to the south and west. Anglers especially enjoy this segment as it slows into a series of still, deep pools, rock banks and gravel bars. In addition to public land access along the river, the Division of Wildlife offers a fishing easement just downstream from Howard, on the river's east side. (Please obey posted rules).

Vallie Bridge to Parkdale
A few miles below Vallie Bridge, the river enters the Lower Arkansas River Canyon, also called the "Grand Canyon of the Arkansas." Towering red-orange cliffs form a spectacular backdrop, bordered for the most part by public lands. This a prime area for wildlife viewing. Bighorn sheep come down the steep slopes to drink from the river. The river drops more sharply in this segment. Rapids with names like Maytag, Lose-your-lunch, Three Rocks and Sharks Tooth make this prime territory for whitewater rafting. Anglers can take advantage of groups of large rocks placed in the river on public lands from Coaldale to Parkdale. These rocks are part of a successful trout habitat improvement project, providing resting places where trout congregate to take a breather from fast currents. Public river access is available at several pull-offs and recreation sites.

CONTROLLING AGENCY Colorado State Parks.

INFORMATION Salida Office (719) 539-7289.

MAP REFERENCES San Isabel National Forest map. Benchmark's *Colorado Road & Recreation Atlas:* pp. 87, 99-101.

Arvada/Blunn Reservoir

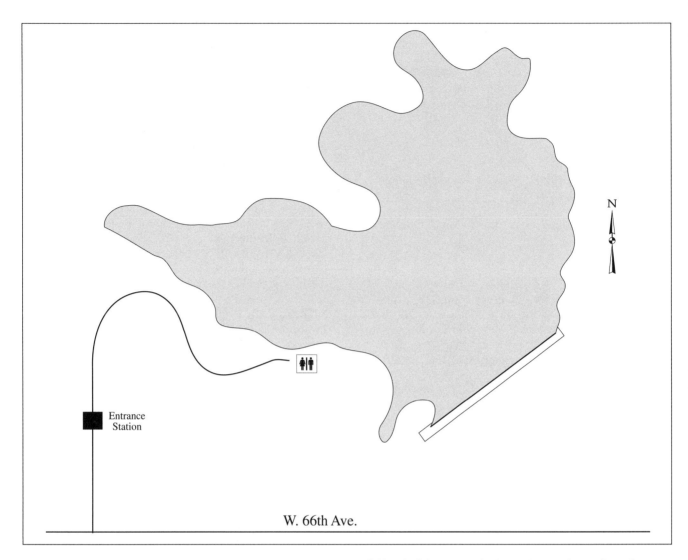

Entrance
Station

W. 66th Ave.

N

DIRECTIONS The reservoir is located between Colorado 93 and Indiana Street on West 64th Ave. in Arvada.

FEE Daily or annual pass.

SIZE 180 surface acres.

ELEVATION 5,758 feet.

MAXIMUM DEPTH 77 feet.

FACILITIES Restrooms accessible for the handcapped.

BOAT RAMP One concrete ramp. Only non-motorized boats are allowed on the reservoir.

FISH Rainbow trout, brown trout, largemouth and small-mouth bass, walleye, and tiger muskie. (Note: The bag and possession limit for trout is two fish.) Trout; fishing is good from shore on Power Bait, Power Eggs, and night-crawlers. Trolling has produced good results with Tasmanian Devils, Dick Nites and Panther Martin lures. Bass;

fishing is fair to good in the early morning or late afternoon on crank baits, spinner baits and assorted lures such as the Cordel Spot. Tiger muskie; large spoons and lures worked along the shallow shorelines. A special permit is required for fishing. Permits are available at the reservoir entrance gate or at Arvada City Hall. Opens April 1 and closes October 31.

RECREATION Fishing, boating, hiking.

CAMPING No camping is allowed. Arvada Reservoir is a day-use only area. Open from sunrise to sunset.

CONTROLLING AGENCY City of Arvada.

INFORMATION Arvada Parks, Hospitality, and Golf Department (720) 898-7417.

MAP REFERENCES Benchmark's *Colorado Road & Recreation Atlas:* p. 134.

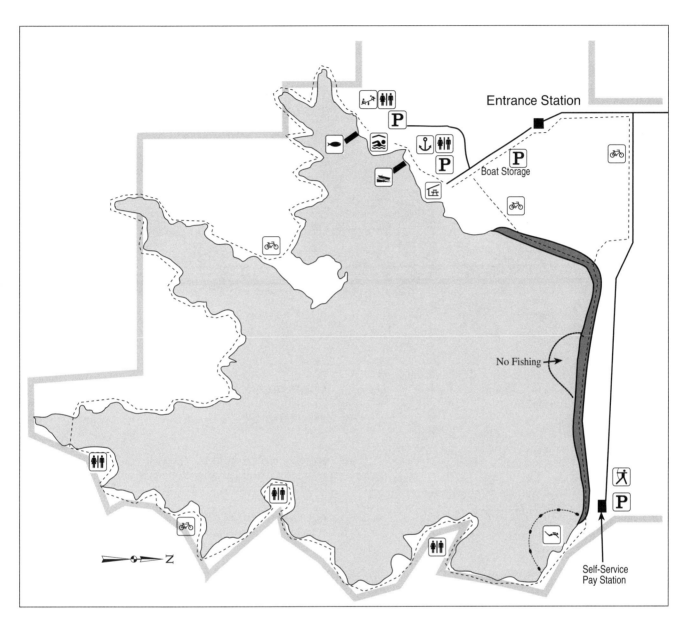

Entrance Station

Boat Storage

No Fishing ←

Self-Service
Pay Station

N

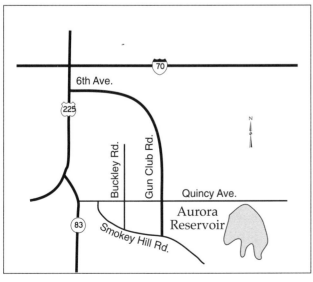

6th Ave.

70

225

Buckley Rd.

Gun Club Rd.

Quincy Ave.

Aurora
Reservoir

Smokey Hill Rd.

83

DIRECTIONS From Parker Road to Quincy Ave., head east on Quincy Ave. The entrance to Aurora Reservoir is 2.5 miles east of the corner of Quincy Ave. and Gun Club Road at 5800 S. Powhattan Road.

FEE City of Aurora daily or annual vehicle pass.

SIZE 820 acres.

ELEVATION 5,930 feet.

MAXIMUM DEPTH 110 feet.

FACILITIES Marina, dry boat storage, store, picnic shelters, restrooms, archery range, boat rental, swim beach, playground, 8.5 mile bike path, snakc area, volleyball, and AWQUA Lounge educational facility.

Aurora Reservoir

BOAT RAMP A four-lane concrete boat ramp is available. No gas-powered boats are allowed, only electric hand or sail powered boats. Gas engines must be in raised position and portable tanks removed.

FISH Fishing is good for rainbow and brown trout, crappie, wiper, catfish, walleye, bass and perch. Check regulations on bag, possession and size regulations.

RECREATION Fishing, boating, sailing, swimming, biking, hiking, volleyball, and scuba diving.

CAMPING No overnight camping, day-use area only.

Open a half-hour before sunrise to one hour after sunset.

CONTROLLING AGENCY City of Aurora.

INFORMATION Parks and Open Space Department (303) 690-1286.

HANDICAPPED ACCESS Handicapped accessible restrooms, picnic tables, fishing pier and trail. Some facilities available with assistance.

MAP REFERENCES Lake bottom topographic map at store. Benchmark's *Colorado Road & Recreation Atlas:* p. 135.

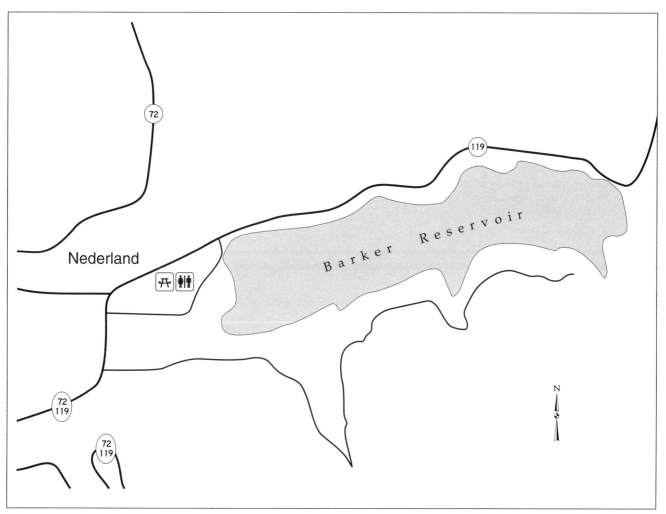

DIRECTIONS   Located immediately east of Nederland on Colorado 119.

FEE No.

SIZE 185 acres.

ELEVATION 8,183 feet.

MAXIMUM DEPTH 132 feet.

FACILITIES Picnic tables, grills and restrooms on north and west sides of the reservoir.

BOAT RAMP Boating and ice fishing are prohibited.

FISH Rainbow, brown and splake trout. Because Barker Reservoir is open for shore fishing only, most of the shoreline offers fairly easy access. Typical baits such as nightcrawlers, salmon eggs and Power Bait usually are your best choices. The best fishing is in the early mornings and rainbow trout generally are the most catchable, along with an occasional brown and splake. Barker Reservoir can receive heavy fishing pressure at times.

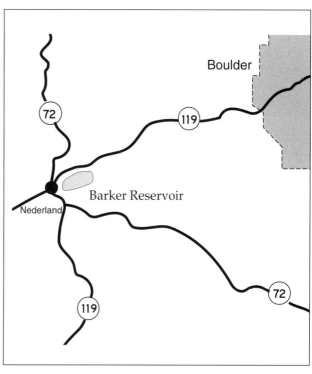

Barker Reservoir

RECREATION Fishing.

CAMPING Picnicking is available at the reservoir in the day-use area but overnight camping is not allowed. However, Kelly Dahl Campground is located 3 miles south of Nederland on Colorado 119. It has 46 campsites, tables, water, and restrooms. This campground is managed by the US Forest Service, there are camping fees and no reservations.

CONTROLLING AGENCY City of Boulder.

INFORMATION Water Resources (303) 258-3266.

MAP REFERENCES Benchmark's *Colorado Road & Recreation Atlas:* p. 62.

Barr Lake State Park

DIRECTIONS From Denver drive north on I-76 to Bromley Lane. Head east on Bromley Lane to Picadilly Road, then south on Picadilly Road to the Barr Lake State Park entrance. Parking areas are on the east side of the lake.

FEE Daily or annual pass.

SIZE 1,950 surface acres.

ELEVATION 5,100 feet.

MAXIMUM DEPTH 42 feet.

FACILITIES Restrooms, picnic tables, duck blinds, nature center.

BOAT RAMP One concrete ramp is available on the northeast side. Note special restrictions for boating: 10-horsepower motor or less, and boating is allowed only on the northern half of the reservoir. The southern half of the reservoir is a wildlife refuge and nature area.

FISH Channel catfish, bass, walleye, bluegill, wiper, tiger muskie and catchable rainbow trout. Trout should hit on typical baits like salmon eggs, Power Bait or trolling with flashers rigged with a nightcrawler or assorted lures. Bass action on typical crankbaits and spinner baits. Cat fishing is best using stink baits, Crawler Links or chicken liver fished on the bottom. Tiger muskies start hitting as they move into more shallow water in search of small bait fish.

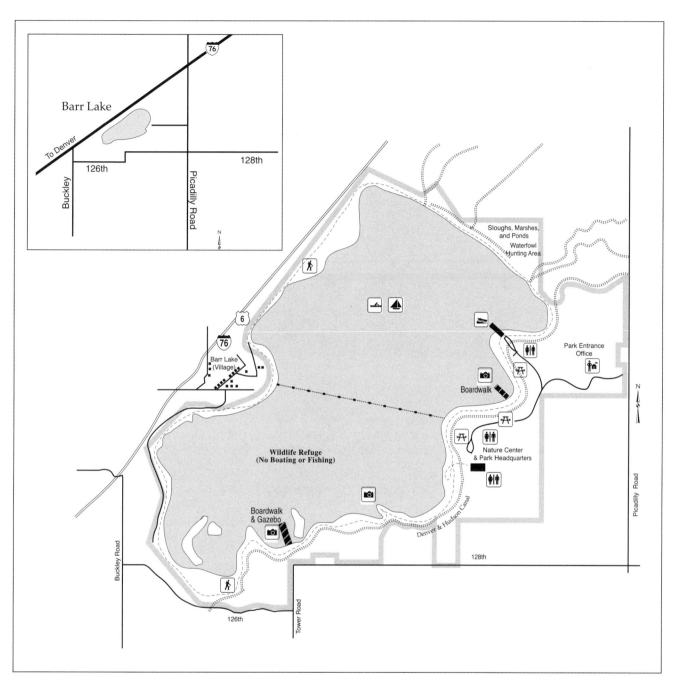

Good action can be expected on larger spoons and top water lures. Barr Lake has extreme seasonal water fluctuation.

RECREATION Fishing, hiking, hunting, birding, biking, horsebackriding, ice fishing, winter activities.

CAMPING No camping allowed. Day use area only; summer hours: open from 5am to 10pm.

CONTROLLING AGENCY Colorado State Parks.

INFORMATION Brighton Office, (303) 659-6005.

MAP REFERENCES Benchmark's *Colorado Road & Recreation Atlas:* p. 63.

OTHER INFORMATION Barr Lake, northeast of Denver, has been a resort area since the mid-1880s. At one time it was an elite outing area for sportsmen and was known as the "finest fishing area in the West." Later, pollution almost ruined the lake, and it has only recently been restored by pollution control projects on the South Platte River and by natural recovery processes. Barr Lake, a major prairie reservoir, was originally made for purposes of irrigation. Its shores are lined with stands of cottonwood, marshes and aquatic plants. The lake forms the heart of 2,715 acre

Barr Lake State Park

Barr Lake State Park, which opened to the public in April 1977. It is administered by the Colorado Division of Parks and Outdoor Recreation.

The southern part of the lake is designated as a wildlife refuge, sheltering fox, deer and other animals and a great number and variety of birds including white pelicans, great blue herons, cormorants, egrets, ducks, grebes, gulls, owls, bald eagles, falcons and hawks.

Recreation at Barr Lake takes forms that harmonize with nature such as hiking, horseback riding, picnicking and nature study. Water activities include sailing, rowing and fishing. In winter, visitors may participate in cross country skiing and snowshoeing.

The park has 3 parking lots. Visitors should drive and park in designated areas only. A boat ramp is near the north lot. No swimming or wading is allowed at Barr Lake and all pets must be leashed. Boating, however, is permitted (except in the refuge area) as long as visitors use only sailboats, hand propelled craft and boats with electric and gasoline motors of ten horse power or less.

Since the lake is stocked by the Colorado Division of Wildlife, anglers will find Barr Lake a good place to fish for channel catfish, bass, walleye, crappie, bluegill, rainbow trout, tiger muskie and wipers. Fishing from the dam is prohibited.

Visitors enjoy the various displays and public programs available at the park nature center. The center provides information on the varied flora and fauna of the area and includes live animal displays and hands-on activities. Also a part of the nature center is a small bookstore offering field guides, educatioal materials, posters and postcards for sale. Hours at the nature center vary.

A nine-mile trail for hikers and bicyclists circles the lake. Shorter walks may be made to the boardwalks that extend over the lake. Visitors who remain on the trails and use insect repellent should have no difficulty with the ants and mosquitoes that form an important part of the food chain for park wildlife. A nature trail begins near the south parking lot and leads to an excellent view of the heron rookery. There are benches along the trail so visitors may rest. Park rangers will conduct bird walks or nature walks for school or youth groups, naturalists or other interested parties. Arrangements should be made in advance by calling the park office. Visitors may ride their own horses on park trails but horses are not permitted on the boardwalks.

The park serves as the headquarters for the Colorado Bird Observatory. Their staff offers public programs and are a valuable source of information about birds. They can be reached at their office, (303) 659-4348.

HANDICAPPED ACCESS Handicapped accessible visitors center, restrooms, picnicking, fishing and hunting facilities. Some facilities available with assistance.

HUNTING Only waterfowl hunting is permitted and is controlled. Check-in is required. Trapping is not permitted. One must contact the park office located at the nature center for regulations.

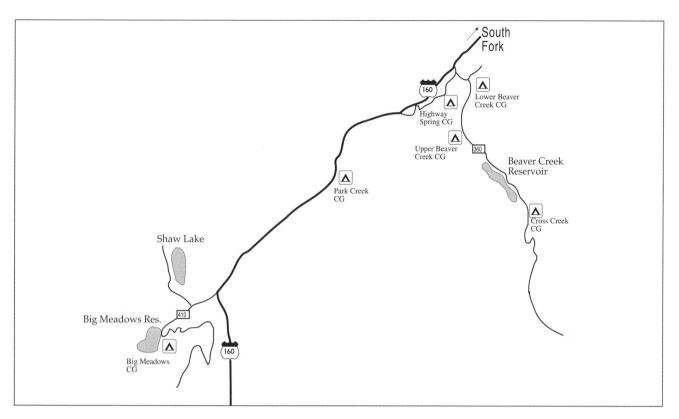

BEAVER CREEK RESERVOIR

DIRECTIONS From South Fork, go 2 miles southwest on US 160 to Forest Road 360 (Beaver Creek Road), then 6 miles to reservoir.

FEE Camping fee.

SIZE 114 acres.

ELEVATION 8,850 feet.

FACILITIES Campground (USFS), drinking water, restrooms, boat ramps.

BOAT RAMP Ramp located near campground. Wakeless boating.

FISH The reservoir is rated good for kokanee, rainbow and brown trout.

RECREATION Camping, boating, fishing, hunting.

CAMPING Cross Creek Campground, 8 sites for units up to 25 feet. Campground has water, restrooms, and fire grates. Beaver Creek Campground, 23 sites for units up to 35 feet.

CONTROLLING AGENCY Rio Grande National Forest. Divide Ranger District (719) 657-3321.

MAP REFERENCES Rio Grande National Forest. Benchmark's *Colorado Road & Recreation Atlas:* p. 112.

BIG MEADOWS RESERVOIR

DIRECTIONS From South Fork, go 12.5 miles southwest on US 160 to Forest Road 410, then 1.8 miles west to reservoir.

FEE Camping fee.

SIZE 114 acres.

ELEVATION 9,200 feet.

FACILITIES Campground (USFS), drinking water, restroom, boat ramp, fishing pier.

BOAT RAMP Ramp located near campground.

FISH Rainbow and brook trout.

RECREATION Camping, boating, fishing.

CAMPING Big Meadows Campground, 60 sites for units up to 35 feet. Campground has water, restrooms, and fire grates.

CONTROLLING AGENCY Rio Grande National Forest. Divide Ranger District. (719) 657-3321.

HANDICAPPED ACCESS Handicapped accessible fishing and trail. Some facilities accessible with assistance.

MAP REFERENCES Benchmark's *Colorado Road & Recreation Atlas:* p. 112.

Big Creek Lakes

DIRECTIONS From Fort Collins, take Colorado 14 west to Walden. From Walden, go north on Colorado 125 nine miles to Cowdrey. Take County Road 6W northwest to Pearl. Just south of Pearl, take Forest Road 600 southwest for six miles to the lower lake. The upper lake requires a two-mile hike from the campground.

FEE Camping fee.

SIZE Lower 343 acres. Upper 101 acres.

ELEVATION Lower–8,997 feet. Upper–9,010 feet.

MAXIMUM DEPTH 58 feet.

FACILITIES Drinking water, vault toilets, picnic tables.

BOAT RAMP Two concrete ramps available, but they can be difficult to use depending on the water level. No motors permitted in Upper Lake.

FISH Lower lake: Brown, brook, rainbow trout, mackinaw, kokanee, grayling, and tiger muskie. Upper lake: Brown, brook, rainbow and mackinaw. (Special restrictions on mackinaw, check fishing regulations). Best results for trout fishing from shore is by using salmon eggs, Power Bait and nightcrawlers on the bottom in the early morning or early evening. Suggested fly patterns include Damselfly and Damsel Nymphs, Matuda, Hornbergs and Woolly Buggers, best fished wet about 4 feet below the surface. Dry patterns include Adams, Royal Coachman and Caddis dressed well on the surface. Mackinaw are caught in deep water using typical large spoons jigged with or without sucker meat. Tiger muskies are best caught on large metal lures such as Panthers, Daredevils and Mepps. Five miles of stream fishing is available in the national forest.

RECREATION Fishing, boating, and hiking. A 1.5 mile self-guided interpretive trail. Big Creek Falls is located two miles from the upper lake, a hike of nearly four miles from the campground area.

CAMPING 54 campsites are available at the US Forest

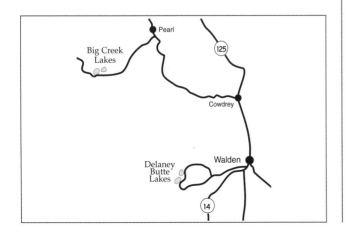

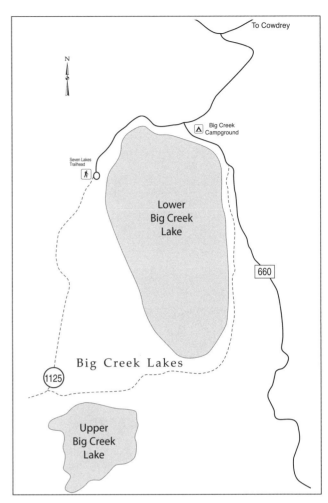

Service campground at the lower lake. Accommodates units to 45 feet. Reservations accepted.

CONTROLLING AGENCY Medicine Bow & Routt National Forests. Parks Ranger District. (970) 723-8204.

MAP REFERENCES Benchmark's *Colorado Road & Recreation Atlas:* p. 46.

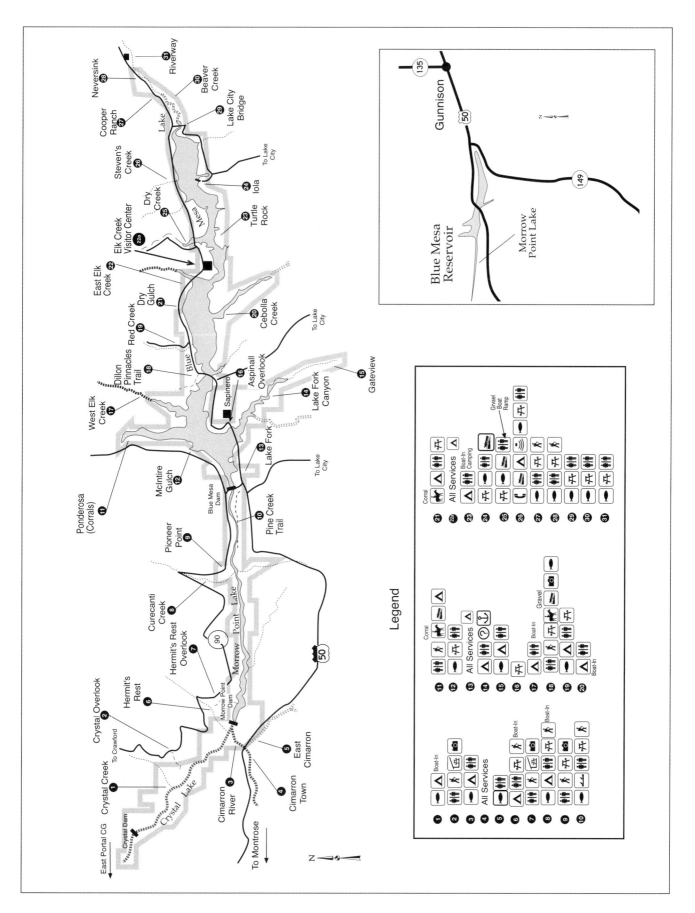

Blue Mesa Reservoir

DIRECTIONS From Pueblo, go west on US 50 to Gunnison. Blue Mesa Reservoir is 10 miles west of Gunnison on US 50.

FEE No entrance fee, except at East Portal. Boating and camping fees.

SIZE Blue Mesa–9,000 acres; Morrow Point–820 acres.

ELEVATION 7,519 feet.

MAXIMUM DEPTH 338 feet.

FACILITIES Most areas offer drinking water. There are three RV sewage dumps. All are accessible to travel trailers.

Lake Fork and Elk Creek marinas offer boat rental and slip rental. Next to Elk Creek marina is Pappy's Restaurant that is open for breakfast, lunch , and dinner.

BOAT RAMP North side; Elk Creek and Stevens Creek. South side; Iola and Lake Fork.

FISH Rainbow trout, perch, sucker, kokanee, brown trout and mackinaw. Rainbow trout, the best action from shore is using bait on the bottom such as nightcrawlers or Power Bait. Trolling from a boat using cow bells with a nightcrawler also produces fair to good rainbow action. Kokanee; the best results for snagging is around the dam and up the Lake Fork of the Gunnison. Trolling flashers, with a night crawler or Cherry Bobber lure are also successful. The best areas for trolling have been the East Iola Basin and Soap Creek area by the dam. Brown trout; good results from boat drift fishing and casting flies near the shallow water areas. The best colors are brown and orange, olive and orange or black in eighth and quarter-ounce sizes. Mackinaw; the best area is the Sapinero Basin near the Sapinero Island. Jigging three-eights-ounce jigheads trailing a piece of sucker meat.

RECREATION Fishing, boating, personal watercraft, wind surfing, swimming, scuba diving, sailing, waterskiing, boat tours.

CAMPING Curecanti offers a variety of drive-in, boat-in, and hike-in campgrounds. Facilities range from Elk Creek Campground with showers, marina, restaurant, amphitheater and visitor center, to the remote West Elk Creek Boat-In Campsite located on the secluded West Elk Arm of Blue Mesa Lake. **Major Campgrounds:** Elk Creek, Lake Fork, Stevens Creek and Cimarron. Combined there are 324 sites in the four campgrounds. Full service campgrounds except there are no electrical hookups or laundries except for some sites in Elk Creek that have electrical hookups. Stevens Creek does not have pull through sites or showers. Cimarron does not have showers. **Limited Facilities Camping**: Dry Gulch, Red Creek, Gateview, Ponderosa, East Portal, and East Elk Creek. All have picnic tables, fire grates, vault toilets and water. Ponderosa has a boat ramp and horse corral.

CONTROLLING AGENCY National Park Service.

INFORMATION Curecanti Recreation Area (970) 641-2337.

SPECIAL RESTRICTIONS Morrow Point Lake is primarily accessible only by the Pine Creek Trail, and private boats must be hand carried. Pets must be kept under physical control. Dispose of waste water only at proper dump sites. Fishing requires a Colorado License. Boaters should be aware of frequent high winds. Head for shore immediately when winds come up.

Blue Mesa Lake, the focus of water sports, is Colorado's largest lake when filled to capacity. Morrow Point and Crystal Lakes, narrow and deep within the canyon carved out by the Gunnison, suggest fjords. Boat tours offered in season on Morrow Point Lake provide an insight into the ancient sculpting work of time and the river.

Curecanti, named for the Ute Chief Curicata who once hunted the Colorado territory provides year round water recreation. The area is managed by the National Park service. The main season for camping, boating, fishing, sailing and sightseeing runs from mid-May until mid-October. One million people come to Blue Mesa Lake every year to fish for Kokanee salmon, rainbow, brown, and Mackinaw trout. During May and early June, brown trout give anglers an opportunity to catch some of the largest fish in Blue Mesa Lake. Occasionally large Mackinaw are also taken then. But, rainbow trout is the mainstay, summer and winter. Trollers find the rainbow and Kokanee holding up well throughout the season, with late June, July, and August best for salmon. October brings spawning runs of Kokanee salmon up the Gunnison River within the recreation area. Hundreds of thousands of rainbow and Kokanee are planted annually in the lake by state and federal wildlife agencies.

Winter offers ice fishing, snowmobiling, cross-country skiing, snowshoeing, and wildlife observation from mid-December through mid-March. Hunting is permitted under federal and state laws. Rock climbers, scuba divers and hang glider enthusiasts pursue their sports here, but if you plan to attempt these, contact a ranger first.

HANDICAPPED ACCESS Handicapped accessible facilities are available in the major campgrounds. Call park office for information.

MAP REFERENCES Benchmark's *Colorado Road & Recreation Atlas:* pp. 96-97.

Spillway

Division of Wildlife Office

Division of Wildlife Residence

Nature Trail

North Cove

Dam Tender's Residence

East Beach

Test Site Point

Ski Counter-Clockwise
Inside Buoys

Entrance Station

Pike's Point

Wagon Wheel Campground

Visitors Center

West Beach

Foster Grove Campground

Park Office

Park Manager's Residence

South Fork Republican

Creek

Landsman

U.S. Highway 385

N

Scale

0   ¼   ½   ¾   1 mile

Bonny Lake

DIRECTIONS Take I-70 east from Denver to the Burlington exit. Travel north out of Burlington on US 385 for 23 miles. Take CR 2 or 3 east 1.5 miles to the Bonny Reservoir area.

FEE None.

SIZE The park covers 1,300 acres.

ELEVATION 3,670 feet.

FACILITIES Picnic tables, a group picnic site with covered pavilion, restrooms, horseshoe pits and cooking facilities.

RECREATION OHV riding, hiking, horseback riding, hunting.

CAMPING Four campgrounds with 190 sites. Sites within each campground can accommodate tents, motor homes or trailers. It is illegal to dump waste and sewage anywhere else. There are individual sites for picnicking as well as a group picnic area.

CONTROLLING AGENCY At present, Bonny Lake is being managed as part of the South Republican State Wildlife Area, a 13,000-acre management unit

INFORMATION Southeast Regional Office in Colorado Springs at 719-227-5200.

OTHER INFORMATION Bonny was a popular migratory waterfowl hunting area during waterfowl season. Bonny Lake State Park is known to be a rest stop for some 70,000

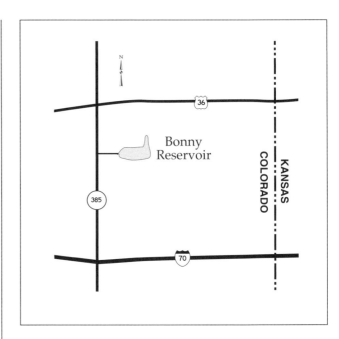

birds representing 250 species during migrations. Between thirty and fifty thousand birds winter there. Among these are snow and Canada geese, wood ducks and osprey as well as prairie falcons, golden and bald eagles. Animal inhabitants of the area include mule and white tail deer, rabbits, coyotes, badgers, muskrats, bobcats and beaver.

During hunting season, all legal methods of hunting may be used, unless otherwise posted. Colorado Division of Parks and Outdoor Recreation and Colorado Division of

Wildlife regulations apply here. Detailed information about the regulations may be obtained from those departments.

CURRENT STATUS   In September 2011, the process of draining Bonny Reservoir began, casting the fate of the park into uncertainty. The action was a result of a 2003 United States Supreme Court ruling in response to a complaint filed by the State of Kansas in 1998. Kansas claimed the neighboring State of Nebraska was in violation of the Republican River Compact, an agreement originally signed in 1942.

The States of Kansas, Nebraska, and Colorado, as well as a representative of the U.S. President negotiated the Republican River Compact. The purpose of the agreement was to equally share the waters of the Republican River drainage, to divide the responsibility of efficient use of those waters as well as effects of destructive floods, and to determine the most efficient use of those waters, and to likewise remove controversy and promote interstate cordiality. Colorado was a party in both the compact and the lawsuit because the headwaters of the Republican River originate in the state's northeastern high plains.

Ultimately, the decision resulted in Colorado owing Kansas billions of gallons of water. Colorado State officials proposed several courses of action to remedy the deficit, including the construction of a new pipeline, but the measures were nonviable. Officials could find no other alternative to draining Bonny Reservoir.

Although the lake was drained, current plans are to keep the park open, and the land has been converted into a State Wildlife Area. Colorado Parks and Wildlife is working with the United States Bureau of Reclamation, Yuma County Commissioners, the Three Rivers Alliance, Yuma County Economic Development Council and other community groups to determine the future of the former State Park. The discussions include facilities and campgrounds that were previously managed by Colorado State Parks. At present, Bonny Lake is being managed as part of the South Republican State Wildlife Area, a 13,000-acre management unit that offers deer, turkey, waterfowl and small game hunting.

During the transition process Bonny Lake remains open to the public. The Visitor Center and Camper Service buildings are closed and water and electricity are not available.

The Colorado State Parks website will post updates on the park's status or enquiries can be directed to the Southeast Regional Office in Colorado Springs at 719-227-5200.

MAP REFERENCES   Benchmark's *Colorado Road & Recreation Atlas:* p. 80.

Boulder Reservoir

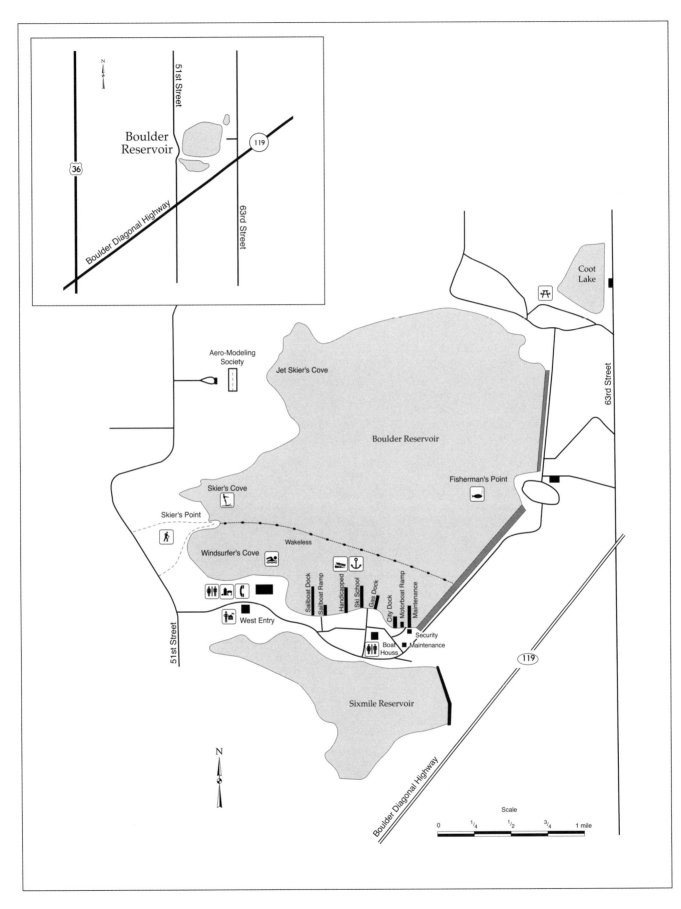

Boulder Reservoir

Coot Lake

51st Street

63rd Street

Boulder Diagonal Highway

119

36

Aero-Modeling Society

Jet Skier's Cove

Skier's Cove

Skier's Point

Windsurfer's Cove

Wakeless

Fisherman's Point

Sailboat Dock

Sailboat Ramp

Handicapped

Ski School

Gas Dock

City Dock

Motorboat Ramp

Maintenance

West Entry

Security

Maintenance

Boat House

Sixmile Reservoir

N

Scale

0   1/4   1/2   3/4   1 mile

Boulder Reservoir

DIRECTIONS From Boulder head towards Longmont on Colorado 119 to 51st Street, then north to the entrance.

FEE Daily or annual pass.

SIZE 700 acres.

ELEVATION 5,173 feet.

MAXIMUM DEPTH 28 feet.

FACILITIES Picnic tables, restrooms, boat rental, marina, food concessions.

BOAT RAMP There is a power boat and sailboat ramp, as well as docks for various water craft.

FISH Trout, perch, bass, bluegill, walleye, crappie, catfish. Fishing rated good. No fishing along south shoreline Memorial Day to Labor Day; anywhere else is fine. Ice fishing weather permitting.

RECREATION Boating, fishing, personal watercraft, wind sailing, waterskiing.

CAMPING No overnight camping. Day-use only. Summer hours 6 am to 7 pm daily. Open year round.

CONTROLLING AGENCY City of Boulder Department of Parks.

INFORMATION Gate (303) 441-3468.

SPECIAL RESTRICTIONS All boats must purchase a daily or season permit to use to lake. No daily permits are sold on weekends or holidays.

MAP REFERENCES Benchmark's *Colorado Road & Recreation Atlas:* p. 63.

Boyd Lake State Park

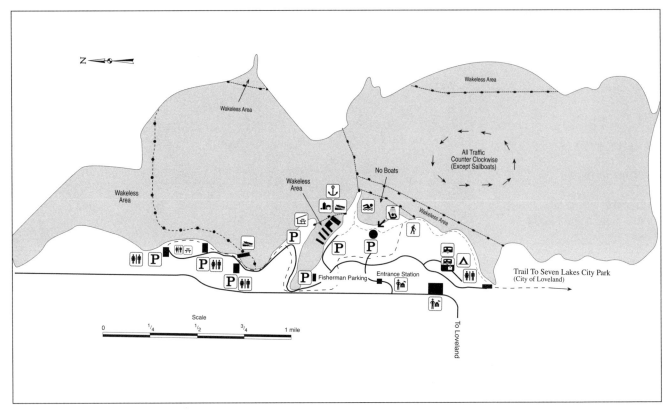

DIRECTIONS From Denver, take I-25 north to the Loveland exit, US 34. Travel west on US 34 into Loveland to Madison Street. Travel north on Madison to County Road 24E, then east on County Road 24E to County Road 11C. Take County Road 11C north to the lake entrance.

FEE Daily or annual pass.

SIZE 1,674 acres.

ELEVATION 4,958 feet.

MAXIMUM DEPTH 50 feet.

FACILITIES A pavilion at the beach supplies showers, restrooms, first-aid station and food concession. There are 95 picnic sites on the west side of the lake. There is also a marina and visitor center.

BOAT RAMP Boating is restricted to certain sections of the lake. There is a paved boat ramp north of the swim beach at the inlet and a paved ramp north of the group picnic area. The Deep Water Ramp is available when the water level is low. The Boyd Lake Marina offers boat rentals, mooring services, fuel, snacks and boating and fishing supplies.

A 6-lane paved boat ramp with docks is located north of the swim beach at the inlet area and a 2 lane paved ramp is located north of the group picnic area.

FISH Boyd Lake is popular for its warm-water fishing for

bass, catfish, brown trout, rainbow trout, bluegill, crappie, walleye and perch, but spring walleye fishing is its main attraction. Expore the ridges on the lake bottom at the south end of the lake or at the inlet cove. Some trout caught will weigh 3-5 pounds. Rainbow trout; using bait from shore or trolling. Shore fishing for trout is good spring, summer and fall. Walleye; with jigs and crank bait.

A fish-cleaning station is located at the restroom building above the boat ramp.

RECREATION Swim beach, boating, biking, hiking, fishing, personal watercraft, waterskiing, sailing, hunting, camping.

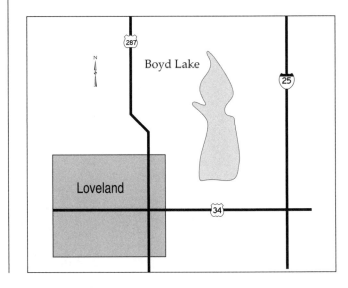

CAMPING There are 148 paved, pull-through campsites with picnic tables, fire pits, electrical hookups, restrooms and dump station. The sites are located on a grassy knoll dotted with trees near the lake. The sites can accommodate tents, pick-up campers, trailers and motor homes. Three restrooms with showers, playground equipment and horseshoe pits are scattered throughout the campground. Reservations are not required but are recommended.

OTHER INFORMATION Boyd Lake State Park is a water sports haven for northern Colorado. The park attracts visitors who enjoy boating, waterskiing, swimming, fishing and windsurfing. The park also features camping and pic-

nicking. Boating is the primary activity on the two-mile-long lake.

The spacious and comfortable 70-degree water of Boyd Lake attracts water skiers of all abilities. Only the south end of the lake is open for waterskiing and the ski pattern is counterclockwise. The entire lake is open to boating and sailing.

The Boyd Lake swim beach is a gentle sloping cove of clean, fine sand, which provides cool relief on Colorado's beautiful summer days. A pavilion at the beach supplies showers, restrooms, a first-aid station and a food concession.

Inner tubes, air mattresses and similar devices may be used in the buoyed-off swim area. Picnic tables are conveniently located just a frisbee throw from the beach.

CONTROLLING AGENCY Colorado State Parks.

INFORMATION Loveland Office (970) 669-1739.

HANDICAPPED ACCESS Handicapped accessible visitor center, restrooms, showers, swimming, trails, fishing and camping. Some facilities available with assistance.

MAP REFERENCES Benchmark's *Colorado Road & Recreation Atlas:* p. 63.

Carter Lake

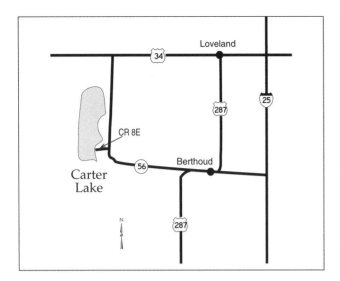

DIRECTIONS From Denver, take I-25 north to the Berthoud exit (Colorado 56). Travel west on Colorado 56 through Berthoud. The highway will take a northern turn, then turn west again. At that point, the road will take you along the east shoreline.

FEE Daily or annual pass.

SIZE 1,100 surface acres.

ELEVATION 5,781 feet.

MAXIMUM DEPTH 180 feet.

FACILITIES Marina, restrooms, water, dump station, and restaurant.

BOAT RAMP Multi-lane concrete boat ramp available on the north end at the marina, South Shore Marina and at North Pines.

FISH Trout, kokanee salmon, walleye, and largemouth bass. Trout; fishing the shore along the dam areas, using salmon eggs and pink Power Bait. Boat fishing trolling flashers with half a nightcrawler, gold medal or Rapala lures. Salmon; trolling with Dick Nites and Needlefish lures. Perch; minnows along the west shore and in the

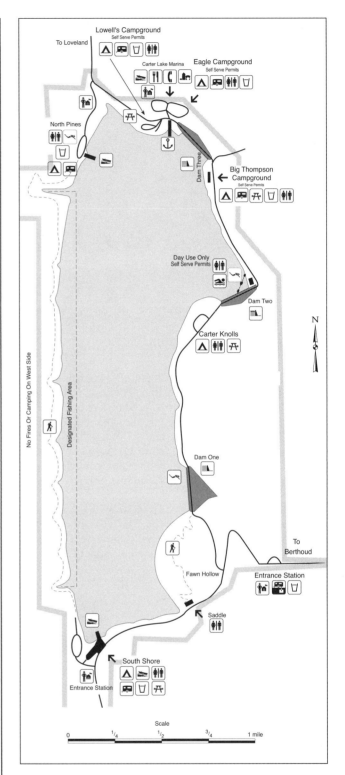

coves. Largemouth bass; in the cove areas on assorted crankbaits and jigs.

Sampling has shown some fish from the water exceeded the mercury action level of 0.5 parts per million. Information regarding mercury concentrations in fish and fish consumption advisories in Colorado can be found at http://www.cdphe.state.co.us/wq/FishCon/Analyses.

RECREATION Sailing, waterskiing, personal water craft, fishing, boating, swimming, wind surfing.

CAMPING Five campgrounds with more than 100 campsites. Most sites have picnic tables, and fire rings. Overnight camping fee. November to March is first-come, first-served. Camping units to 40 feet.

HANDICAPPED ACCESS Handicapped accessible restrooms located at South Shore and North Pines boat launching areas. Eagle and Big Thompson Campgrounds.

SPECIAL RESTRICTIONS Quiet hours 10 PM to 6 AM. No fires on the west side of the lake. Swim beach is located at Dam #2. Swimming is prohibited elsewhere. Cliff diving and jumping into the water is prohibited.

INFORMATION Carter Lake Office (970) 679-4570.

OTHER INFORMATION Carter Lake is part of the Colorado-Big Thompson Project which is operated by the United States Bureau of Reclamation and the Northern Colorado Water Conservancy District.

Carter Lake Reservoir is one of the two main project storage reservoirs in the east slope distribution system. It is filled by the pumping unit at Flatiron Power Plant which pumps from Flatiron Afterbay through a connecting pressure tunnel 1 1/3 mile long. The pumping lift through this tunnel varies from a minimum of 162 feet to a maximum of 286 feet, depending on the water surface elevation in the reservoir. During peak power demands on the project system, the flow through this tunnel can be reversed and the pumping unit at Flatiron used as a generator.

The reservoir as constructed has a total capacity of 112,000 acre-feet of which approximately 109,000 acre-feet is active or usable. The partly natural site provides the cheapest storage per acre-foot of any project reservoir. The outlet, which is in the 285-foot-high main dam on the southeast side of the reservoir, has a capacity of 625 cubic feet per second.

In addition to the main dam, there are two smaller dikes across low saddles in the surrounding hills. Construction of Carter Lake Reservoir began in July 1950 and was completed in October 1952.

The reservoir is managed by the Larimer County Parks and Open Lands Department. Carter Lake Reservoir is popular for fishing, sailing, camping and waterskiing. The park is open year-round.

CONTROLLING AGENCY Larimer County Parks and Open Lands Department.

MAP REFERENCES Free Larimer County Parks Map. Benchmark's *Colorado Road & Recreation Atlas:* p. 63.

Catamount & Crystal Creek Reservoirs

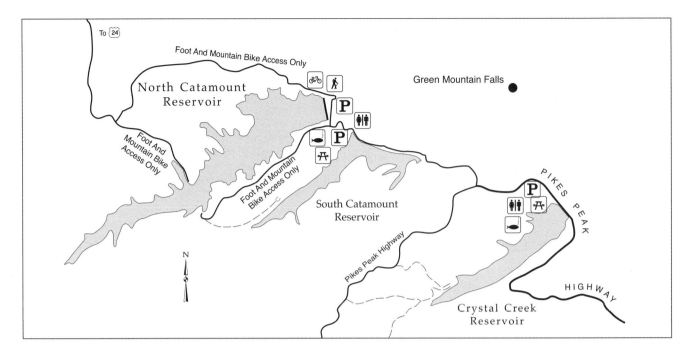

DIRECTIONS From Colorado Springs, travel west on US 24 to the town of Cascade. From Cascade drive south on Pikes Peak Highway to the reservoirs.

FEE Daily fee and fee to access Pikes Peak Highway.

SIZE North Catamount—210 surface acres, South Catamount—120 acres, Crystal Creek—136 acres.

ELEVATION 9,344 feet.

MAXIMUM DEPTH Crystal—50 feet at spillway, South Catamount—65 feet at spillway, North Catamount—120 feet-deep to 135 feet at the spillway.

FACILITIES Restrooms, picnic tables, fishing pier.

BOAT RAMP No ramps, but hand-launch boats, including belly boats are allowed with hand-powered or electric motors only. Sail boating is not permitted. Boats must be brought to the area on car top carriers or in the back of pick-up trucks. Trailers are not allowed on Pikes Peak Highway.

FISH Cutthroat, brook, rainbow and mackinaw at North and South Catamount. Brook, rainbow and cutthroat trout at Crystal Creek. North Catamount is open for flies and lures only. The others allow all methods. Areas open for bait fishing should produce on salmon eggs, cheese bait, nightcrawlers or Power Bait.

Areas restricted to flies and lures will produce good results on fly and bubble combos with Renegades, Bloody Butchers, Mosquitoes, Adams, and Pistol Petes. Hardware fishermen can expect good action on Tasmanian Devils, Kastmasters, Thomas Buoyant and Roughrider lures. North Catamount has a good population of brookies and a good population of mackinaw with a minimum size limit of 20 inches. All reservoirs are limited to a four trout bag limit that can include no more than one mackinaw. The best fishing is near the dam or in the deeper water coves.

RECREATION Fishing, mountain biking on designated trails, hiking, wildlife viewing, picnicking, and scenic photography.

CAMPING No camping is allowed. Day use area only.

CONTROLLING AGENCY City of Colorado Springs and Division of Wildlife.

INFORMATION North Slope Recreation Area (719) 684-9138.

HANDICAPPED ACCESS Handicapped accessible parking lots, restrooms and a fishing pier accessed by a concrete ramps.

HOURS May 1–June 1, 9 AM–4 PM; May 23–September 1, 7 AM–8:30 PM; September 2–September 30, Saturday and Sunday, 7 AM–4 PM and Monday through Friday, 8 AM–4 PM.

MAP REFERENCES Benchmark's *Colorado Road & Recreation Atlas:* p. 89.

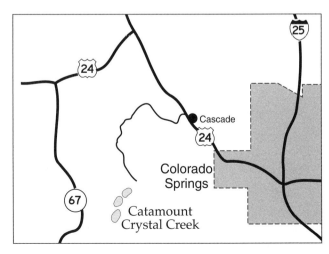

Chapman Reservoir

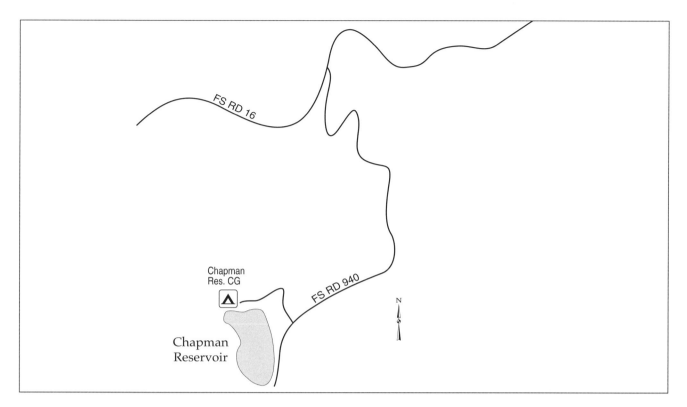

DIRECTIONS From Denver, travel west on I-70 to the town of Wolcott. At Wolcott, travel north on Colorado 131 toward Steamboat Springs to Yampa. At Yampa, travel west on Forest Road 16 (Flattops Scenic Byway) towards Dunckley Pass. After about 10 miles turn south onto Forest Road 940. Travel about one mile to Chapman Reservoir.

An access road to Sheriff's Reservoir is 2 miles farther west on Forest Road 16, then south on Forest Road 959/960. Sheriff's Reservoir is 25 acres, rated fair to good for 10-inch rainbow. Chapman and Sheriff reservoirs are clearly marked by signs along Forest Road 16.

FEE Camping fee.

SIZE 37 surface areas.

ELEVATION 9,280 feet.

FACILITIES Vault toilet, fire rings, designated sites, parking lot.

BOAT RAMP There is no boat ramp, and boating is restricted to hand-power or electric trolling motors only.

FISH Stocked rainbow, cutthroat and brook trout. Shore fishing; salmon eggs, Power Bait, nightcrawlers or a combination will produce good action early or late in the day. Flyfishing; Damsel Nymphs, Matuka, Hornbergs and Woolly Bugger patterns fished west about 3 feet below the surface with a casting bubble. Hardware fishing; Thomas Cyclone, Thomas Buoyants, Mepps and Rooster Tail lures.

RECREATION Fishing, hunting, OHV riding, camping, ice fishing, snowmobiling.

CAMPING 12 developed sites at Chapman Reservoir Campground. No-fee camping available along Forest Road 16.

CONTROLLING AGENCY Medicine Bow & Routt National Forests.

INFORMATION Yampa Ranger District (970) 638-4516.

MAP REFERENCES Routt National Forest map. Benchmark's *Colorado Road & Recreation Atlas:* p. 59.

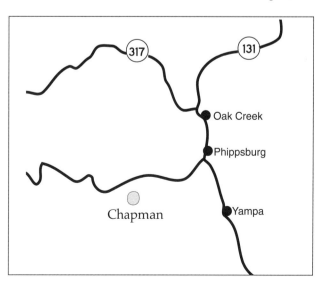

Chatfield State Park

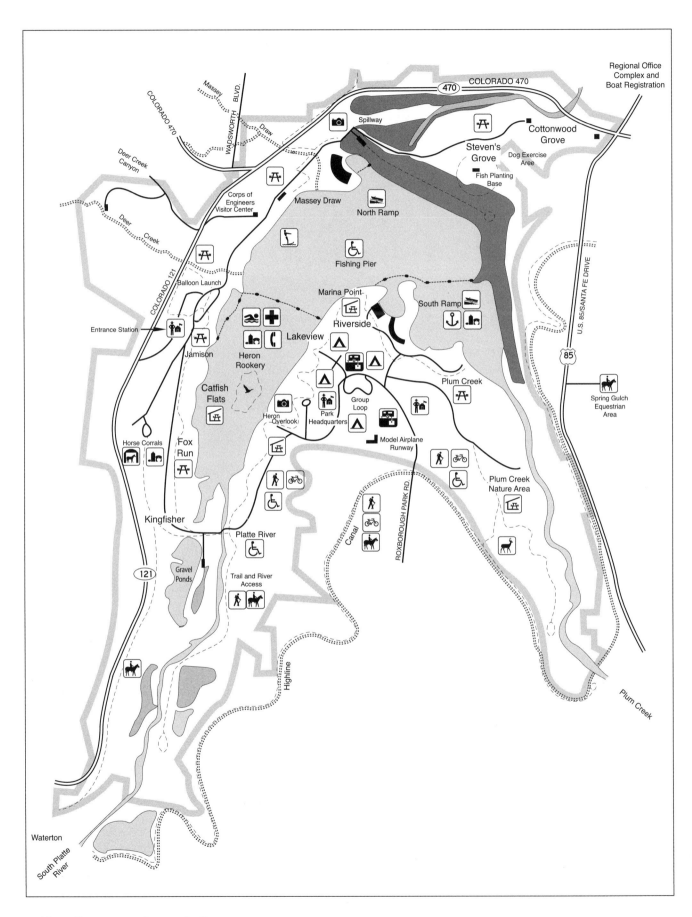

Chatfield State Park

DIRECTIONS From Denver, travel south on Wadsworth Blvd. After you pass under C-470, continue south on Colorado 121. Turn east into the Deer Creek entrance of the Chatfield Reservoir State Park.

FEE Daily or annual pass.

SIZE 1,450 surface acres.

ELEVATION 5,432 feet.

MAXIMUM DEPTH 60-70 feet.

FACILITIES Complete service park. Dump station, laundry, showers, group camping areas, group picnic areas, bathhouse, swim beach, trails, snack bar, marina, boat rentals and Jet Ski rentals.

BOAT RAMP The north boat ramps have eight launch lanes and four courtesy docks. The south boat ramp has two launch lanes and one courtesy dock. There are two large No Wake Zones at the southwest and southeast ends of the reservoir.

FISH Trout, bass, perch, bluegill, sunfish, carp, crappie, walleye and catfish. Trout; Power Bait or salmon eggs from shore near the northwest area. Trolling for trout on Cow Bells and Rapalla, Kastmaster lures, nightcrawlers or minnows. Walleye; jigs and minnows. Some bass are being caught on surface baits in the evenings. Catfish; stink baits at night near the Catfish Flats and Plum Creek inlet areas.

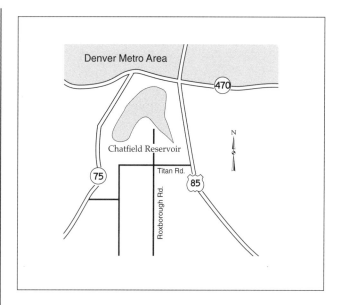

RECREATION Boating, sailing, hot air ballooning, model airplane flying, personal watercraft, swimming, hiking, biking and horseback trails.

CAMPING Four camping areas provide 197 sites. All are available on a reservation basis. Facilities include pull-through or back-in campsites, flush toilets, hot showers, laundry, centrally located water, and firewood. 120 sites have water, electrical, and sewage hookup and 77 have electrical hookup. Ten group campsites are available on a

Chatfield State Park

reservation basis only. Campers may stay a maximum of 14 days in any 45-day period. Camping allowed only in designated areas. A holding tank dump station is located near the campground. Quiet hours are from 10 PM to 6 AM. Individual sites can be reserved.

CONTROLLING AGENCY Colorado State Parks.

INFORMATION Littleton Office (303) 791-7275.

HANDICAPPED ACCESS Chatfield State Park is fully accessible to persons with disabilities. Twenty-five miles of paved trails are wide enough and level enough to accommodate wheelchairs. A handicapped fishing pier is located near the marina on the east side of the lake, and an access trail is located at the South Platte River. Accessible visitors center, showers, restrooms, swimming. Some facilities available with assistance.

SPECIAL RESTRICTIONS No hunting or discharge of firearms at any time. No access is permitted into the Great Blue Heron nesting area, March 1 to September 1.

OTHER INFORMATION Chatfield Reservoir is part of the 5,600 acre Chatfield State Park. Located on the prairie near the foothills southwest of Denver, it is the state of Colorado's second busiest recreation area hosting some 1,500,000 visitors annually.

Although there are numerous other activities available, waterskiing, sailing, fishing, camping and swimming are the primary attractions. Visitors are required to adhere to zoning that regulates boating, swimming, and waterskiing.

There are two entrances to the park: the Deer Creek Entrance on the northwest side, off Colorado 121/470, and one on the south side, off Titan Road. The Army Corps of Engineers runs a visitor center which provides information on the dam construction as well as on the flood control concept and operations.

The boat ramp is just north of the 121/470 entrance. The swimming area is on the northwest side. Picnic sites with tables and grills are located throughout the park. Many of the group picnic areas are sheltered.

A diversity of other activities not found at most lakes or reservoirs is available at Chatfield. There are approximately 25 miles of paved bicycle paths in the park. A nature study area is located on the south side of the park along Plum Creek. An observation area provides an excellent view of a 27-acre heronry, a large, protected great blue heron nesting area. No humans are allowed in this area during the nesting season. Guided nature walks and evening programs are given if staffing permits.

The hundreds of acres of water, prairie, grasses and ground cover provide a wildlife preserve for mule and whitetail deer, hawks, eagles, owls, shorebirds, waterfowl, beaver, muskrats, coyotes, rabbits and other small mammals. No hunting, shooting of firearms or archery is allowed anywhere in the park area.

MAP REFERENCES Free park map.
Benchmark's *Colorado Road & Recreation Atlas:* p. 75.

Cheesman Lake

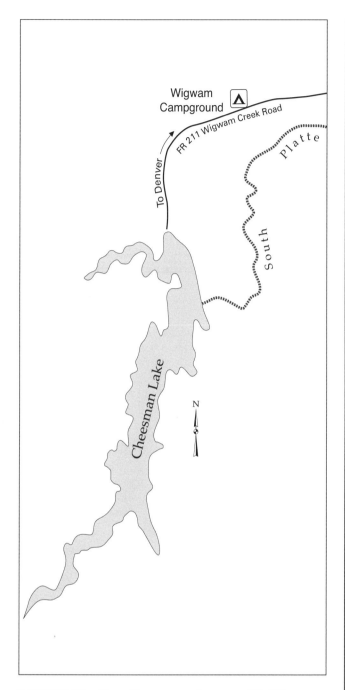

DIRECTIONS   From Denver travel west on US 285 to Pine Junction. Drive south to Pine on County Road 126, about 20 miles to Forest Road 211. Go west 1 mile then the road turns south, at which point it is 1.5 miles farther to the lake.

FEE No.

SIZE 875 surface acres.

ELEVATION 7,382 feet.

MAXIMUM DEPTH 190 feet.

FACILITIES Restrooms.

BOAT RAMP None, shore fishing only—no boats.

FISH Rainbow, brown, and brook trout, northern pike, sucker, kokanee salmon, yellow perch. Fishing by artificial flies or lures only. Possession and size limit for trout is two fish that are 16 inches or longer. Fishing is prohibited from January through to the end of April. No night fishing. Kokanee snagging permitted from September 1 to December 31.

RECREATION Fishing.

CAMPING No camping. Picnicking is permitted throughout the park.

CONTROLLING AGENCY Denver Water Department.

INFORMATION Colorado Division of Wildlife, Colorado Springs Office (719) 227-5200.

OTHER INFORMATION In 2002, the Hayman fire destroyed much of the forest surrounding Cheesman Lake. The forest service has estimated that it will take 150 years for the forest to re-establish itself. However, vegatation and groundcover has already started growing in the area.

MAP REFERENCES Pike National Forest map. Benchmark's *Colorado Road & Recreation Atlas:* p. 88.

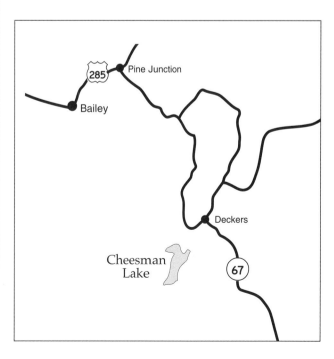

Cherry Creek State Park

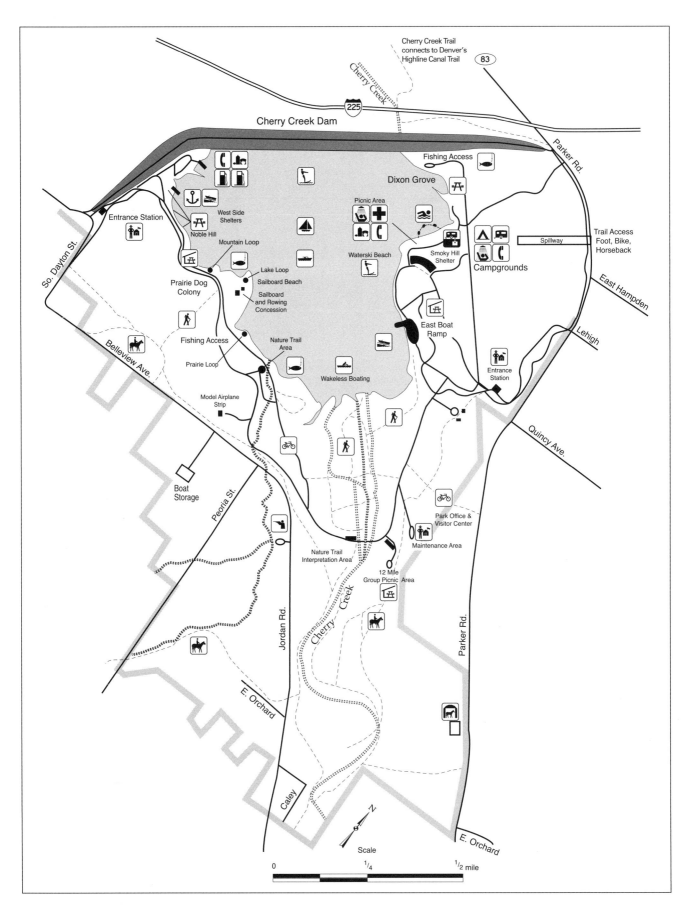

Cherry Creek Trail connects to Denver's Highline Canal Trail

83

Cherry Creek

225

Cherry Creek Dam

Parker Rd.

Fishing Access

Dixon Grove

Entrance Station

So. Dayton St.

West Side Shelters

Noble Hill

Mountain Loop

Picnic Area

Smoky Hill Shelter

Spillway

Trail Access Foot, Bike, Horseback

East Hampden

Campgrounds

Prairie Dog Colony

Lake Loop

Sailboard Beach

Sailboard and Rowing Concession

Waterski Beach

Lehigh

Fishing Access

Nature Trail Area

Prairie Loop

East Boat Ramp

Belleview Ave.

Model Airplane Strip

Wakeless Boating

Entrance Station

Boat Storage

Peoria St.

Jordan Rd.

Quincy Ave.

Park Office & Visitor Center

Maintenance Area

Nature Trail Interpretation Area

12 Mile Group Picnic Area

Parker Rd.

Cherry Creek

E. Orchard

Caley

N

Scale

0   1/4   1/2 mile

E. Orchard

Cherry Creek State Park

DIRECTIONS Take I-225 to Parker Road then head south to the east entrance. West entrance is off S. Yosemite Street.

FEE Daily and annual pass.

SIZE Reservoir—880 surface acres; Park—4,200 acres.

ELEVATION 5,550 feet.

MAXIMUM DEPTH 26 feet.

FACILITIES Trails, model airplane field, shooting range, stables, swimbeach. A holding tank dump station is across from the campground entrance. There are also boat rental and marina facilities.

BOAT RAMP There are two boat ramps, one on the east shore and one on the west shore.

CAMPING 125 newly renovated sites with showers and laundry plus a group camping area. There is a limited number of electrical hookups. No more than six people may occupy each site. Larger groups may occupy additional sites. Maximum stay is 14 nights in a 45-day period. Camping reservations 1-800-678-2267 (Outside Denver); (303) 470-1144 (Denver area); Group Reservations Call (303) 690-1166.

FISH Bass, bluegill, crappie, perch, pike, channel catfish, trout, tiger muskie, walleye, wiper, carp. Handicapped accessible fishing pier. Cherry Creek has the state record for walleye with many large walleye caught. Best time is in April and May at night along the dam.

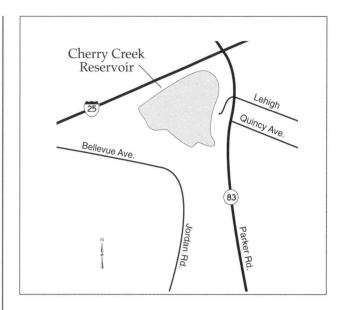

RECREATION Fishing, waterskiing, jet skiing (rental at Ski Shack), sailing and boating plus a beach for swimming. Swimming is allowed only in designated areas. Depending on weather conditions, winter recreation includes ice boating, snowmobiling, sledding, ice fishing, ice skating and cross-country skiing.

Hunting is prohibited throughout Cherry Creek State Park. There is, however, a rifle range with fixed targets for rifles and pistols and a trap area for shotguns. There are three self-guided nature trails for interpretive walks, as well as a prairie dog observation area. Ranger led walks are by reservation. The campground has an amphitheater and Saturday evening programs.

Cherry Creek State Park

Other recreation opportunities include 12 miles of paved bicycle trail circling the reservoir; a model airplane field with asphalt runways, frequency posts, field regulations for radio controlled aircraft and 12 miles of horse trails and a horse stable with horses for rent all year long.

There is also a dog training area, seasonal food and bait concession, and seasonal first-aid station at the beach.

CONTROLLING AGENCY Colorado State Parks.

INFORMATION Aurora Office (303) 690-1166.

OTHER INFORMATION Running through what was once rolling ranch country and paralleling portions of the historic Smoky Hill Trail, Cherry Creek flooded on several occasions and brought destruction to Denver. To control Cherry Creek, the US Army Corps of Engineers completed Cherry Creek Dam in 1953, creating Cherry Creek Reservoir, which soon became a popular recreation resource. Through a long-term lease negotiated between the state and the Army Corps of Engineers, the area around the reservoir became Cherry Creek State Recreation Area in 1959.

Many improvements have increased recreation opportunities for the one-and-a-half million annual visitors. For picnicking there are many lake shore sites with tables and grills. There are five group reservation picnic areas that accommodate 250 people each; some have shelters.

SPECIAL RESTRICTIONS Vehicles must remain on designated roads or in designated parking areas at all times. A valid parks pass is required at all times, and a campground permit if you are camping. Dogs must be kept on a leash at all times. Fires are permitted only in grills or provided fire rings. Do not leave personal belongings unattended on picnic tables or in tents. Lock you car. Do not tie any lines or ropes to the trees or other vegetation. Gathering firewood is prohibited. Park gates close at 10 PM and open at 5 AM daily. The park often reaches capacity on summer weekends.

HANDICAPPED ACCESS Special facilities to accommodate handicapped persons are located at the fishing access areas, Dixon Campground, the group picnic sites and the swim beach. Parking for handicapped visitors is signed throughout the park. Accessible showers, restrooms, swimming, picnic tables, trails, fishing, some facilities available with assistance.

MAP REFERENCES Free park map.
Benchmark's *Colorado Road & Recreation Atlas*: p. 75.

Chicago Lakes

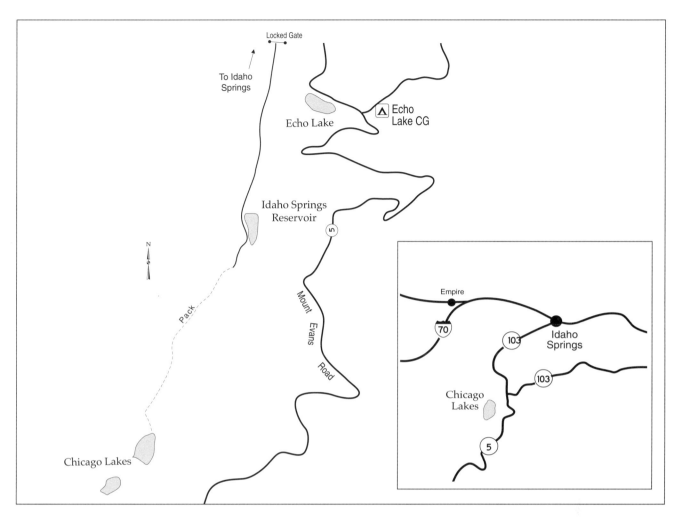

DIRECTIONS From Denver, travel west on I-70 to Idaho Springs. From Idaho Springs, travel south on US 103 about 8 miles (past the Chicago Forks Picnic Ground) to Chicago Creek Road. Travel south 1.5 miles on Chicago Creek Road to Idaho Springs Reservoir, where you will begin about a 2 mile hike. Hike around the west side of Idaho Springs Reservoir and follow the feeder creek upstream. The hike is about 2 miles south to Chicago Lakes.

FEE No.

SIZE Upper lake—10 acres; lower lake—26 acres.

ELEVATION Ranges from 11,400 feet to 11,700 feet.

MAXIMUM DEPTH Upper: 41 feet; Lower: 74 feet.

FACILITIES None.

BOAT RAMP No ramp. No boating allowed.

FISH Cutthroat trout. The area is known for fair-to-good cutthroat trout fishing. It might be heavily used by fishermen, campers, hikers, and picnickers in the summer. The best fishing tends to be at the first access in the spring or later in the fall. Good action can be expected using either typical baits, salmon eggs, worms or lure fishing with assorted styles of Mepps or Rooster Tails. The best action on flies, depending on the time of the year, can be on Adams, Blue Dun, Olive Dun for dries and Gold Ribbed Hares Ear for nymphs. If you choose to fish Idaho Springs Reservoir, this body of water has 20 acres, a 30 foot maximum depth. Fishing is for brook and cutthroat trout.

RECREATION Fishing, hiking, picnicking.

CAMPING Open camping. However, special regulation for camping in wilderness area where all campsites must be at least 100 feet from any water source such as lakes and streams.

CONTROLLING AGENCY Arapaho/Roosevelt National Forest.

INFORMATION Clear Creek Ranger District (303) 567-3000.

MAP REFERENCES Benchmark's *Colorado Road & Recreation Atlas:* p. 74.

Colorado State Forest

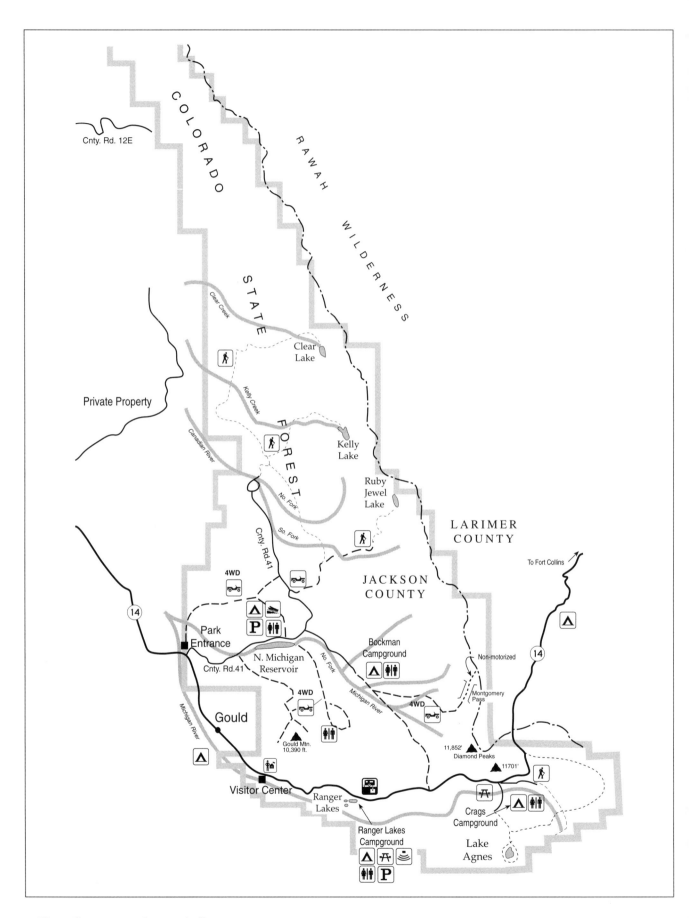

DIRECTIONS From Fort Collins on US 14 travel west up Poudre Canyon and over Cameron Pass. There are three entrances into the forest on the west side of Cameron Pass: Lake Agnes, Ranger Lake, and North Michigan. The Visitor Center and park office is located off of US 14, 1 mile east of Gould.

FEE Daily or annual pass. Camping fee.

SIZE The park covers approximately 70,000 acres, with 170 surface acres of lakes.

ELEVATION Ranges from 8,500 to 12,500 feet, so visitors may need time to acclimate.

BOAT RAMP Boating on North Michigan Reservoir is allowed at wakeless speeds. There is a concrete boat ramp on the north side of the reservoir.

RECREATION Fishing, hunting, camping, hiking, backpacking, horseback riding, wildlife viewing, moose viewing, four-wheel drive and OHV riding, snowmobiling, snowshoeing and cross-country skiing.

CAMPING The State Forest offers four developed campgrounds. Crags and Ranger Lakes can be accessed from US 14, while North Michigan Reservoir and Bockman are accessed from County Road 41. These campgrounds combined offer 158 sites with picnic tables and grills with water available in all but Bockman. The State Park also offers dispersed camping in the areas north of Ruby Jewel Lake Road and south of the Bockman campground. Minimum impact camping techniques are requested. Stop at the park office to learn more about these areas.

CABINS The State Forest has 6 rustic cabins for rent. The cabins include wood burning stoves, bunk beds, and picnic tables with outdoor grills. All these cabins are rented on a reservation system and can be booked online at *http://parks. state.co.us/Parks/StateForest/CabinsandYurts/Cabins.*

The park ranges from plain grassy meadows to heavy black timber. Hiking trails range in skill level from very easy to almost technical climbs. In the early morning, be sure to keep a close watch around the willow covered beaver ponds for Colorado moose.

YURTS Yurts are circular tent-like canvas and wood structures located in remote areas that can be rented for overnight stays in the park. For information, rates and park locations contact Never Summer Nordic at (970) 723-4070.

FISH Native brook, rainbow, brown and cutthroat trout. The high mountain lakes are Clear Lake, Kelly Lake, Ruby Jewel Lake, American Lakes and Lake Agnes. The current state record golden trout, caught in Kelly Lake in 1979, was 3 pounds, 12 ounces and 22.5 inches. Lake Agnes is noted for moderate-sized cutthroat, but several 7-8 pound fish are caught each year. All high-country lakes are re-

stricted to flies and lures. Lures that have worked are the Thomas Ruff Rider, Kastmaster, and an eighth of an ounce or smaller, a small No. 3 or No. 5 Rapala Countdown in gold, or a Thomas Arrow. Steamers are your best patterns, such as Hornberg Supervisor, Carey Special or Halfback and Fullback nymphs. A great number of small feeder streams in the area and North Michigan Reservoir are not restricted. Salmon eggs and nightcrawlers are the best. The best fishing generally is in the early morning or late afternoon near the drop-offs on all lakes.

Ranger Lakes are stocked with rainbow trout, but you can expect to hook a brown or brook trout on occasion. Fishing baits from shore such as pink or yellow Power Bait, salmon eggs or marshmallows. Lures; red or yellow spotted Panther Martins, Jakes, Spin-A-Lure and Mepps Black Fury lures. Flyfishing; hornberg dries, Mosquito, Prince Nymphs, Float-N-Fools, Royal Wulff, Black Halfbacks and Adam's fly patterns. No boating of any kind is allowed on Ranger Lakes. The lower lake is 4.3 surface acres, the upper lake is 8.5 surface acres.

CONTROLLING AGENCY Colorado State Parks.

INFORMATION Walden Office (970) 723-8366.

HANDICAPPED ACCESS Handicapped accessible restrooms, hunting, picnic tables, camping and fishing. At North Michigan Campground, site #16 located near the lake has 2 handicapped-accessible restrooms. Bockman Campground has one handicapped accessible restroom.

MAP REFERENCES Benchmark's *Colorado Road & Recreation Atlas:* p. 47.

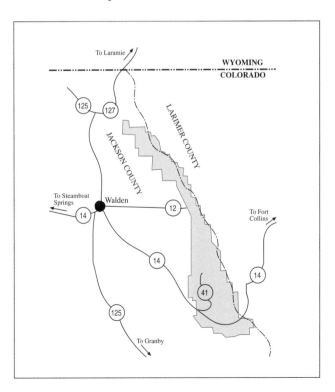

Crawford State Park

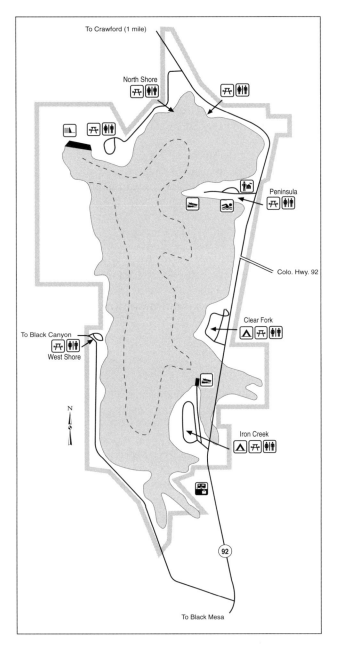

DIRECTIONS From Delta drive 20 miles east on Colorado 92 to Hotchkiss. Then continue on Colorado 92 to Crawford. Crawford State Park is one mile south of Crawford on Colorado 92.

FEE Daily or annual pass. Camping fees.

SIZE 397 surface acres.

ELEVATION 6,553 feet.

MAXIMUM DEPTH 120 feet.

FACILITIES Dump station, restrooms, picnic tables, swim beach, trails, visitor center.

BOAT RAMP One boat ramp is at the end of the peninsula

past the ranger station and the other is on the northeast side of Iron Creek Campground.

CAMPING Two campgrounds with 66 campsites, 45 sites with electric and water hook-ups and 21 without. Drinking water is available close-by. A total of 61 sites can accommodate trailers, campers and motor homes, and there are 5 walk-in tent sites. Tables, grills and use pads are at each campsite. Also, 7 sites are handicapped accessible. Both campgrounds have flush toilets and showers.

FISH Yellow perch, largemouth bass, channel catfish, rainbow trout, black crappie and northern pike.

RECREATION Boating, sailing, jet skiing, personal watercraft, scuba diving, biking, hunting, winter recreation, swimming.

OTHER INFORMATION Crawford State Park offers its visitors camping, fishing, water sports, hunting and numerous other leisure time activities in scenic mountainous terrain. The area around the park is almost exclusively cattle country. As the center of the cattle industry in the North Fork area, the nearby town of Crawford sees hundreds of cows herded down main street on their way either to market or mountain pasture. Ranches and farms still surround the park, drawing water from the same reservoir that affords visitors so many recreational opportunities all year-round. The park's 6,600 ft. elevation guarantees visitors a mild climate at any season.

The famous and spectacular Black Canyon of the Gunnison is only 11 miles from the park. Nearer landmarks visible from Crawford State Park are Needle Rock, Castle Rock and Cathedral Peak.

Within the boundaries of Crawford State Park are 334 land acres and the 400-acre reservoir, which was built in 1963 by the US Bureau of Reclamation. The Colorado Division of Parks and Outdoor Recreation has administered the area since 1965.

INFORMATION Colorado State Parks. Crawford Office (970) 921-5721.

HANDICAPPED ACCESS Handicapped accessible restrooms, swimming, picnic tables, fishing and camping. Some facilities available with assistance.

MAP REFERENCES Free park map.
Benchmark's *Colorado Road & Recreation Atlas:* p. 96.

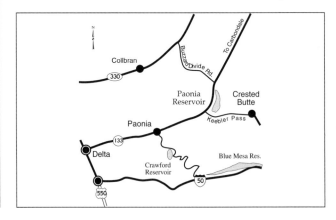

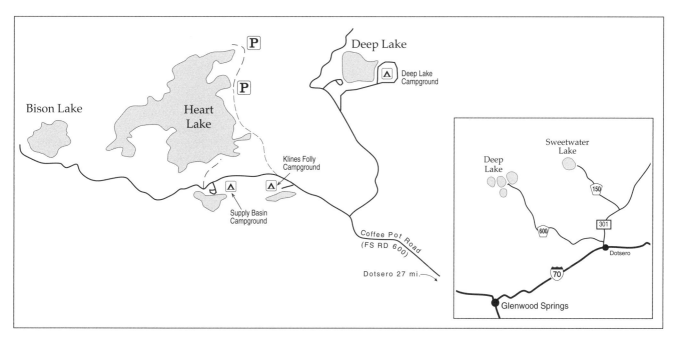

DEEP LAKE

DIRECTIONS Drive 17 miles east of Glenwood Springs on I-70 to the Dotsero exit. Travel north on County Road 301, 1.8 miles, to Forest Road 600 (Coffee Pot Road). Then head northwest on Forest Road 600, 29 miles to Deep Lake.

FEE Camping fees.

SIZE 37 acres.

ELEVATION 10,462 feet.

FACILITIES Campground and boat ramp.

BOAT RAMP There is a good gravel boat ramp in the day-use area for launching of small non-motorized boats with electric motors.

FISH Rainbow, brook and lake trout. Bait fishing from shore with salmon eggs, worms or Power Bait on the bottom generally produces good results. Casting from shore in the early mornings or late evenings with lures, such as Panther Martins, Mepps and Rooster Tails usually will produce good action. The best fishing for lake trout usually is in the spring at ice-off. But lake trout are caught throughout the year by deep water trolling, jigging with large spoons, Gitzits, or lead-headed jigs tipped with sucker meat. Check regulations for size, bag and possession limit.

RECREATION Fishing, hiking, horseback riding.

CAMPING There are 35 sites all with picnic tables and fire rings. There are three vault toilets as well as parking for trailers under 35 feet. No water. Camping is on a first-come, first-served basis. Trash removal is not provided, so please pack it out. Other campgrounds are Klines Folly, with 4 sites, toilet, picnic tables and fire rings, and Supply Basin, which offers 7 sites, fire rings and toilets.

HEART LAKE

DIRECTIONS Heart Lake is located just southwest of Deep Lake. Heart Lake, also governed by the US Forest Service, is rated fair for rainbow and brook trout.

SIZE 480 acres.

ELEVATION 10,700 feet.

BOTH LAKES

CONTROLLING AGENCY White River National Forest.

INFORMATION Eagle Ranger District (970) 328-6388.

OTHER INFORMATION Bank fishing is difficult because of shallow shoreline. Gas motors not permitted.

MAP REFERENCES White River National Forest map. Benchmark's *Colorado Road & Recreation Atlas:* p. 71.

Delaney Butte Lakes

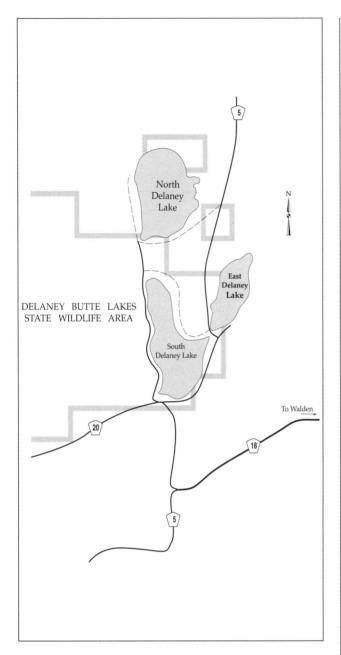

FACILITIES Several shade shelters and picnic tables available. Pit toilets. No drinking water available.

BOAT RAMP One concrete ramp on each lake.

FISH Fishing at all three lakes is good for trout. South Delaney Butte Lake: Rainbow trout and Snake River Cutthroat, average size 16 inches, some up to 22 inches. North Delaney Butte Lake: Brown trout, 60% of Brown in lake over 18 inches, averaging over 3 pounds. Largest recorded Brown trout was 25 inches, 7.5 pounds. North Delaney Butte Lake has slot limit, brown trout between 14- and 20-inches must be returned to the water. East Delaney Butte Lake: Rainbow, Snake River Cutthroat and Brook trout, average size is 13 inches. Known for faster action of the three lakes. Fishing by artificial flies and lures only in all lakes. Flies; olive scud, woolly bugger, caddis emerges and an Adams. Lures; Kastmaster and Mepps. Check for bag, size and possession limits.

RECREATION Fishing, non-motorized boating.

CAMPING Open camping is allowed in the state wildlife area near all three lakes. Pit toilets are available, and some sites offer shaded picnic tables.

CONTROLLING AGENCY Colorado Division of Wildlife.

INFORMATION Fort Collins Office (970) 472-4300.

MAP REFERENCES Routt National Forest map. Benchmark's *Colorado Road & Recreation Atlas:* p. 46.

DIRECTIONS From Denver, go north on I-25 to Colorado 14, exit at Fort Collins, then go west on Colorado 14 to Walden. From Walden, go southwest for 0.5 mile on Colorado 14 to County Road 12W. Take County Road 12W west for 5 miles to County Road 18. Then drive north on County Road 18 to County Road 5. Take County Road 5 for 1 mile north to Delaney Butte Lakes State Wildlife Area.

FEE No.

SIZE East Lake—65.2 acres; North Lake—163.5 acres; South Lake—150 acres.

ELEVATION 8,145 feet.

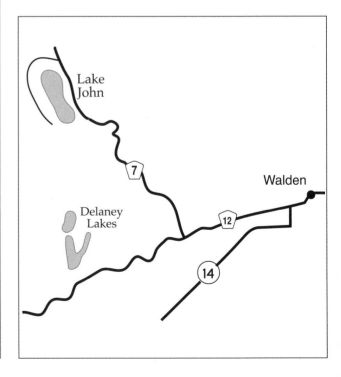

Dillon Reservoir

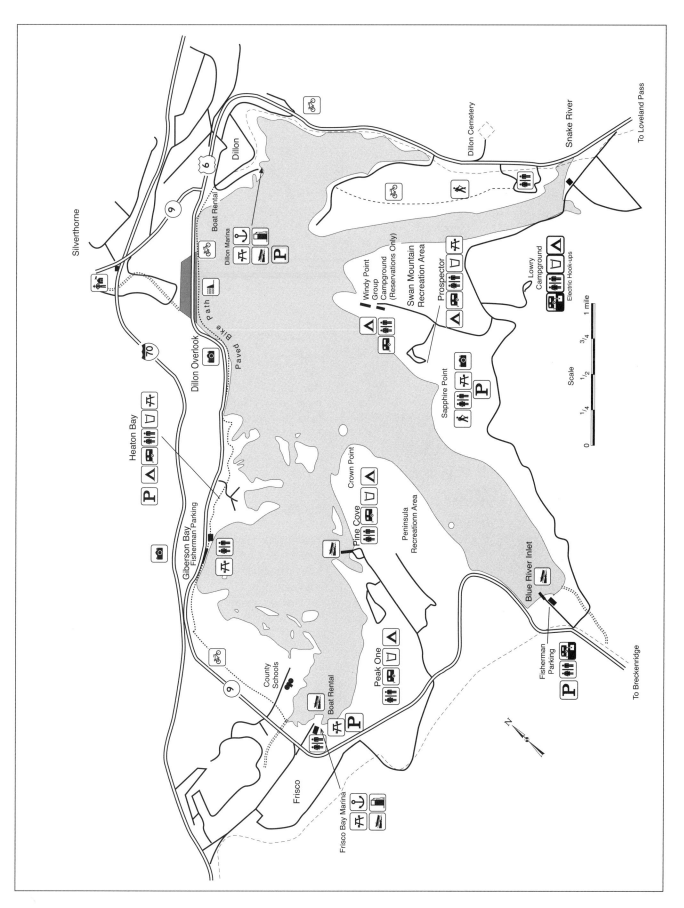

Silverthorne

Dillon

To Loveland Pass

Snake River

Dillon Cemetery

6

9

Boat Rental

Dillon Marina

P

Windy Point Group Campground (Reservations Only)

Swan Mountain Recreation Area

Prospector

Lowry Campground

Electric Hook-ups

70

Dillon Overlook

Paved Bike Path

Sapphire Point

P

Scale

0   1/4   1/2   3/4   1 mile

Heaton Bay

P

Crown Point

Pine Cove

Peninsula Recreationn Area

Blue River Inlet

Giberson Bay Fisherman Parking

9

County Schools

Peak One

Boat Rental

P

Fisherman Parking

P

To Breckenridge

N

Frisco

Frisco Bay Marina

Dillon Reservoir

DIRECTIONS From Denver, travel west on I-70. You can access the reservoir either via the Silverthorne exit or US 6 over Loveland Pass.

FEE Camping fees.

SIZE 3,233 surface acres when full.

ELEVATION 9,000 feet.

MAXIMUM DEPTH 207 feet.

FACILITIES All areas have picnic tables, restrooms, fire grates and drinking water.

BOAT RAMP Boat launching ramps are available at Dillon Marina, Frisco Marina, Blue River inlet and Pine Cove near the Peak 1 Campground. The marinas rent boats and tackle.

FISH Rainbow trout, brown trout, and sucker. The reservoir, which also has a kokanee population, is known for producing large brown trout. Catchable rainbow trout are stocked. Fishing for rainbow is best in late spring and early

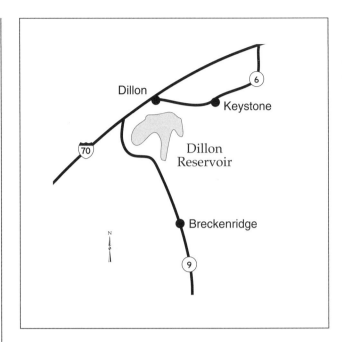

summer using baits such as salmon eggs, cheese baits and worms from shore near any of the feeder streams. It's best from boats trolling pop gear or flashers. The best action on large browns is usually in late evening or during the night, from shore near the feeder stream areas on various baits. Kokanee fishing will range from slow to good, depending on the time of year. General hot spots for shore fishing are near the Blue River inlet, Giberson Bay and Snake River inlet.

RECREATION Sailing, boating, fishing, biking.

CAMPING There are five US Forest Service Campgrounds around the lake on the east, south and west sides. They are all easily reached by Colorado 9, US 6 or Swan Mountain Road.

   Peak 1 Campground has 79 sites, Pine Cove has 55 sites, Heaton Bay has 72 sites, and Prospector has 108. Lowry Campground has 29 sites available on a first-come, first-served basis. Windy Point group site available by reservation only.

   There also are several day-use areas, including the Blue River inlet, Giberson Bay and Snake River inlet. All day-use areas have no facilities and fires are not allowed; there are no fees for day-use.

CONTROLLING AGENCY Denver Water.

INFORMATION White River National Forest, Dillon Ranger District (970) 468-5400.

SPECIAL RESTRICTIONS No water contact sports, such as swimming, waterskiing, wading, or scuba diving are permitted due to the domestic use of the water. Wind surfing is allowed with wet suit.

Dillon Reservoir

OTHER INFORMATION Located high in the Rocky Mountains and backdropped by snowcapped peaks, Dillon Reservoir was built in the 1960s by the Denver Water Department. It is near the towns of Silverthorne, Frisco and Dillon. The original Town of Dillon is at the bottom of the reservoir.

Dillon Reservoir is part of the Denver Water Supply System and covers 3,300 surface acres. Waters feeding into the reservoir are the Blue River, Snake River and Ten Mile Creek.

The first site of the town of Dillon was originally known as "La Bounte's Hole." It was named for a French Canadian trapper and was a rendezvous site for trappers. It was located north of the Snake and the Blue Rivers.

HANDICAPPED ACCESS Handicapped accessible facilities; Prospector Campground located on the east side of the reservoir has handicapped accessible restrooms; Giberson Bay located on the west side of the reservoir near Colorado 9 and I-70 exit has handicapped accessible restrooms and fishing pier; Sapphire Point located south of Prospector Campground has handicapped accessible restrooms and 0.3 miles of paved trail.

MAP REFERENCES White River National Forest map. Benchmark's *Colorado Road & Recreation Atlas:* p. 73.

Dumont Lake

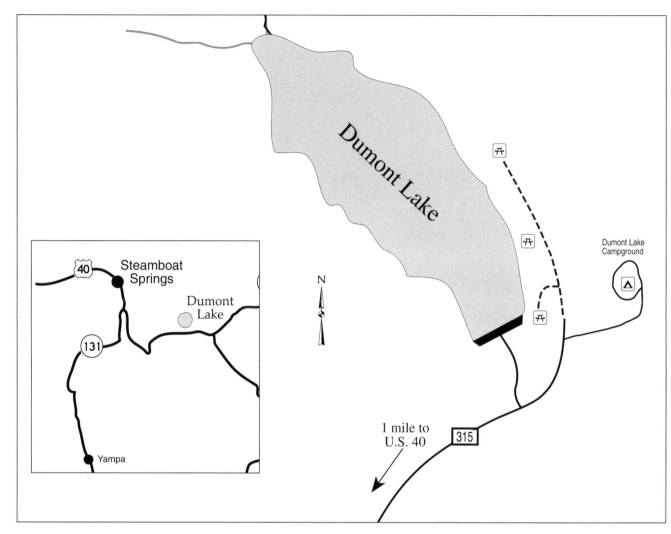

DIRECTIONS From Denver, travel west on I-70 to the Dillon exit. Travel north from Dillon on Colorado 9 to Kremmling. At Kremmling, continue north on US 40 over Muddy Pass and Rabbit Ears Pass. About one mile west of the Rabbit Ears Pass summit, turn north onto Forest Road 315 at a sign for the lake.

FEE Camping fee. Parking fee at picnic area.

SIZE 35 acres.

ELEVATION 9,508 feet.

FACILITIES Water (seasonal), picnic tables, fire rings and vault toilets.

BOAT RAMP No boat ramps. Open for hand-launch only. Restricted to electric or hand-powered craft.

FISH Stocked rainbow trout. A few larger trout are occasionally caught. Fishing is fair to very good using salmon eggs, Power Bait and nightcrawlers from shore. There is good action on small metal lures or small dry fly patterns

in the evenings. The most productive method of fishing is from small boats or belly boats.

RECREATION Hiking, fishing, non-motorized boating, mountain biking, wildflower viewing.

CAMPING Camping is available at Dumont Lake Campground, which has 22 sites, units to 40 feet. Drinking water and toilets at picnic area and campground. No camping is allowed in the picnic area. One drawback with the area is the typical high country summer mosquito population, so go prepared.

CONTROLLING AGENCY Medicine Bow & Routt National Forests.

INFORMATION Hahns Peak/Bears Ears Ranger District (970) 879-1870.

HANDICAPPED ACCESS Campground is handicapped accessible.

MAP REFERENCES Routt National Forest map. Benchmark's *Colorado Road & Recreation Atlas:* p. 60.

Eleven Mile State Park

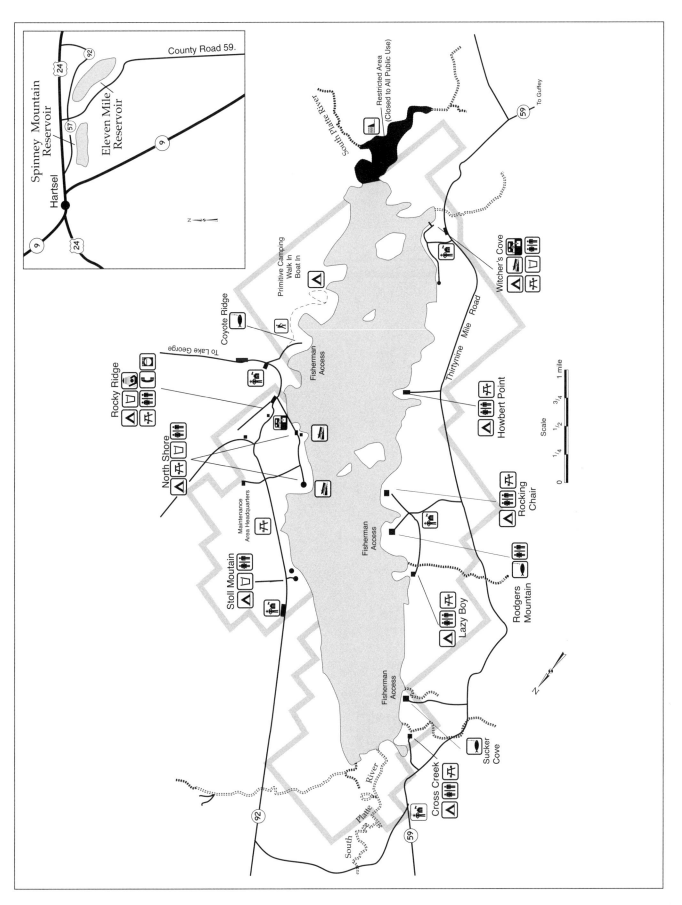

Spinney Mountain Reservoir

Eleven Mile Reservoir

Hartsel

County Road 59.

Restricted Area (Closed to All Public Use)

To Guffey

South Platte River

Primitive Camping Walk In Boat In

Coyote Ridge

Witcher's Cove

Fisherman Access

To Lake George

Thirtynine Mile Road

Rocky Ridge

Howbert Point

North Shore

Scale

0 1/4 1/2 3/4 1 mile

Maintenance Area Headquarters

Rocking Chair

Fisherman Access

Stoll Moutain

Rodgers Mountain

Lazy Boy

Fisherman Access

Sucker Cove

South Platte River

Cross Creek

Eleven Mile State Park

DIRECTIONS From Denver, travel south on I-25 to Colorado Springs, then west on US 24 past the town of Lake George. One mile west of Lake George turn left and follow County Roads 90/92, 10 miles to Eleven Mile State Park.

FEE Daily or annual park pass.

SIZE 3,400 surface acres.

ELEVATION 8,600 feet.

MAXIMUM DEPTH 117 feet.

FACILITIES Picnic tables, fire pits, grills, freshwater pumps, marina, playground, trails, visitor center, vault toilets and dump stations. A camper services building is located at entrance to Rocky Ridge Campground.

BOAT RAMP Three concrete ramps, two at North Shore and one at Witchers Cove.

FISH Anglers have taken some of the largest fish in the state from Eleven Mile's waters. Brown, cutthroat and rainbow trout, carp, kokanee and northern pike abound. Fishing is prohibited in the restricted area at the dam on the east end of the reservoir, but is permitted anywhere else on the reservoir. Bow fishing for carp is permitted year-round. For trout use marshmallows, nightcrawlers, salmon eggs or Power Bait on the bottom. Successful lures to consider are Bunny Jigs and Kastmasters in silver or gold. Trolling produces good results using lures and nightcrawlers. Pike are caught on orange or silver Kastmasters or Rapalla lures. Kokanee have been caught between Duck and Deer Island trolling Cow Bells, Kastmasters, Kokanee Killers or Red Magic lures.

RECREATION Fishing, boating, birding, wind surfing, hunting, picnicking, hiking, and wildlife observation. Winter sports.

SPECIAL RESTRICTIONS No water contact sports, such as swimming, waterskiing, wading, or scuba diving are permitted.

CAMPING Nine campgrounds offer a variety of camping facilities. 300 campsites can accommodate tents, pickup campers and trailers. 25 primitive campsites are located in the back country area at the east end of the reservoir. Holding tank dump stations are located at the entrances to North Shore and Witchers Cove.

OTHER INFORMATION Eleven Mile Reservoir lies to the west of Colorado Springs. Since 1932, when the dam on the South Platte River was completed, it has been an important water storage facility for the City of Denver. The 3,400 acre reservoir forms the center of Eleven Mile State Park (4,000 land acres) under lease from the Denver Water Board. The park acquired its name because the dam was constructed at the entrance to Eleven Mile Canyon. The Pike National Forest borders the park on two sides. In the distance Pikes Peak and the Tarryall Range on the east and the snowy Continental Divide on the west form an imposing setting for the reservoir. The 8,600-foot elevation means that temperatures vary widely in any season. Sunburn can occur rapidly in the thin air.

Boaters (particularly sailors) find their skills fully tested by the tricky winds and fast rising storms that can occur here. Storms can develop quickly at Eleven Mile bringing sudden strong winds and high waves that can be hazardous to small craft. Underwater hazards are frequently found in the reservoir and not all can be marked. Boaters need to be especially careful when within 150 feet of any shoreline. In addition to the Colorado boating statutes and regulations, boaters must observe the following special regulations: (1) Boat docks are for loading and unloading only; mooring at or fishing from the docks is prohibited and a five minute use limit is enforced. (2) The reservoir is closed to boating a half hour after sunset through a half hour before sunrise. (3) All islands are closed to public use.

Waterfowl of many kinds are abundant at the reservoir and in the surrounding area. Antelope, elk, deer, bear, coyote, mountain lion, bobcat and several varieties of small mammals frequent the park. Hunting is permitted during legal seasons in areas not posted as closed. Trapping is by permit only.

Winter recreation enthusiasts come here for ice fishing, ice boating, ice skating and cross country skiing. Snowmobiling is not allowed on the reservoir or within the park unless otherwise posted at the park entrance.

HANDICAPPED ACCESS Facilities for the handicapped include reserved parking spaces, and picnic tables, showers, and campgrounds that have been adapted for their use, some facilities accessible with assistance.

CONTROLLING AGENCY Colorado State Parks.

INFORMATION Lake George Office (719) 748-3401.

MAP REFERENCES Pike National Forest map. Benchmark's *Colorado Road & Recreation Atlas:* p. 88.

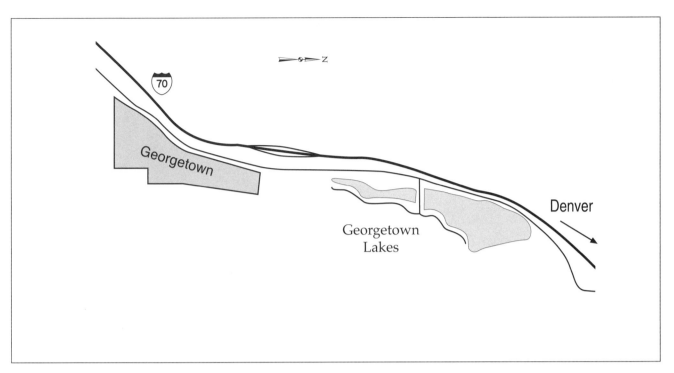

DIRECTIONS From Denver, travel west on I-70 to the Georgetown exit 228. The lakes are located northeast of Georgetown.

FEE No.

SIZE 54 surface acres.

ELEVATION 8,471 feet.

FACILITIES Restrooms.

BOAT RAMP One boat ramp. Boating is allowed but only for hand-launched and hand-powered craft. No motors of any kind allowed.

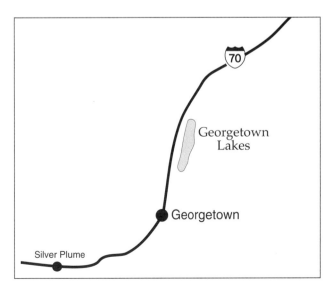

Georgetown Lakes

FISH Stocked with rainbows and brown trout. A good population of brook trout also exists. Typical baits such as salmon eggs, worms or Power Bait, rigged to float just off the bottom, will produce good action for the stocked rainbow and brown trout. Shore fishing with lures using Panther Martins, Roostertails and small gold or silver Kastmasters will also produce good results.

RECREATION Fishing.

CAMPING Camping is restricted by the Town of Georgetown to self-contained units such as campers, trailers and motor homes. Tent camping is not allowed. The closest tent camping facility is at the Clear Lake Campground in the Arapaho/Roosevelt National Forests south of Georgetown. The area also offers a wildlife viewing section for bighorn sheep.

CONTROLLING AGENCY Town of Georgetown.

INFORMATION Georgetown Visitor Center (303) 569-2405.

MAP REFERENCES Benchmark's *Colorado Road & Recreation Atlas:* p. 74.

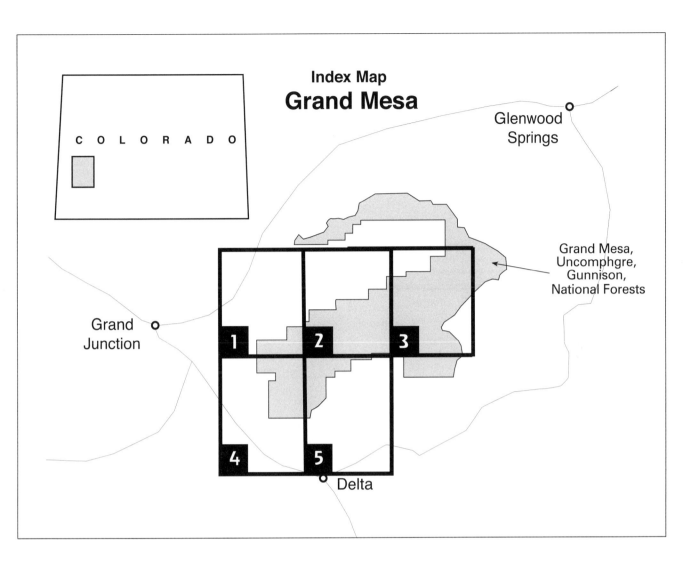

The Grand Mesa is in fact a very large, flat-topped mountain of volcanic origin whose summit area covers some 800 square miles. The Grand Mesa is well known for its numerous natural lakes. In all, some 300 lakes and reservoirs lie on the benches and top of the Grand Mesa. Of these, roughly a third provide the aquatic habitat needed to sustain trout populations on a year-round basis.

The Grand Mesa offers fishing opportunities for seven species of trout including rainbows, Colorado River cutthroat, Snake River cutthroat, brown, and brook trout, as well as splake (hybrid between lake and brook trout), and Arctic grayling. This guide lists 100 lakes and 25 streams, which are monitored or managed by the Colorado Division of Wildlife.

Almost all of the lakes and streams listed lie between 8,000 and 11,000 feet in elevation. Keep in mind your physical abilities as you recreate at these elevations. This elevation combined with strenuous activity can aggravate or trigger health problems.

Access ranges from roads suitable for four-wheel drive or motor homes, to trails accessible by foot and horseback only.

INFORMATION Grand Mesa, Uncomphgre, and Gunnison National Forests, Grand Valley Ranger District (970) 242-8211.

LISTING OF GRAND MESA LAKES AND RESERVOIRS

MAP 1
PAGE 67

KANNAH CREEK BASIN AREA

MAP 2
PAGE 67

MESA LAKES AREA

Jumbo Reservoir
Sunset Lake
Beaver Lake
Glacier Springs Lake
Mesa Lake
Mesa Lake, South
Lost Lake
Mesa Creek
Waterdog Reservoir

BULL BASIN AREA

Griffith Lake #1
Griffith Lake, Middle
Bull Creek Reservoir #1
Bull Creek Reservoir #2
Bull Creek Reservoir #5
Bull Creek

ISLAND LAKE AREA

Island Lake
Little Gem Reservoir
Rim Rock Lake

WARD LAKE AREA

Carp Lake
Ward Lake
Alexander Lake
Hotel Twin Lake
Baron Lake
Deep Slough Reservoir
Ward Creek Reservoir
Ward Creek

EGGLESTON LAKE AREA

Eggleston Lake
Reed Reservoir
Kiser Slough Reservoir
Youngs Creek Reservoir #1
Youngs Creek Reservoir #2
Youngs Creek Reservoir #3
Pedro Reservoir
Kiser Creek
Little Grouse Reservoir
Stell Lake
Kiser Creek
Youngs Creek

CRAG CREST TRAIL AREA

Forrest Lake
Eggleston Lake, Upper, Little
Buffs Lake

TRICKLE PARK AREA

Military Park Reservoir
Stell Lake, East
Park Reservoir
Vela Reservoir
Elk Park Reservoir
Knox Reservoir
Trout Lake

COTTONWOOD LAKES AREA

Silver Lake
Forty-Acre Lake
Neversweat Reservoir
Kitson Reservoir
Cottonwood Reservoir #1
Cottonwood Reservoir #4
Lily Lake
Cottonwood Creek

BONHAM RESERVOIR AREA

Bonham Reservoir
Big Creek Reservoir #1
Atkinson Reservoir
Atkinson Creek
Big Creek

MAP 3
PAGE 68

BONITA RESERVOIR AREA

Cedar Mesa Reservoir
Bonita Reservoir
Trio Reservoir
Bonita Creek

WEIR & JOHNSON RESERVOIR AREA

Twin Lake #1
Twin Lake #2
Sackett Reservoir
Weir and Johnson Reservoir
Leon Peak Reservoir (Sissie Lake)
Finney Cut Lake #1
Finney Cut Lake #2
Cole Reservoir #1
The Pecks Reservoir #1
The Pecks Reservoir #2

LEON CREEK AREA

Rock Lake
Youngs Lake
Kenney Creek Res. (Kendall Res.)

Lost Lake
Kenney Creek
Monument Reservoir #1
Colby Horse Park Reservoir
Leon Lake
Lanning Lake
Hunter Reservoir
Leon Creek
Leon Creek, East
Leon Creek, Middle
Marcott Creek
Monument Creek
Park Creek

LEROUX CREEK AREA

Doughty Reservoir
Hanson Reservoir
Dogfish Reservoir
Goodenough Reservoir
Doughty Creek
Leroux Creek, East
Leroux Creek, West

BUZZARD CREEK AREA

Buzzard Creek
Owens Creek
Willow Creek

MAP 4
PAGE 69

KANNAH CREEK BASIN AREA

Blue Lake
Sheep Creek

MAP 5
PAGE 70

THE DOUGHSPOONS AREA

Carson Lake
Doughspoon Res. #1 (Delta Res. #1)
Doughspoon Res. #2 (Delta Res. #2)
Dugger Reservoir
Morris Reservoir
Porter Reservoir #4 (Little Davies)

THE GRANBYS AREA

Granby Reservoir #1
Granby Reservoir #2
Granby Reservoirs #4, 5, 10, 11
Granby Reservoir #12
Granby Reservoir #7
Battlement Reservoir, Big
Battlement Reservoir, Little
Clear Lake
Dirty George Creek

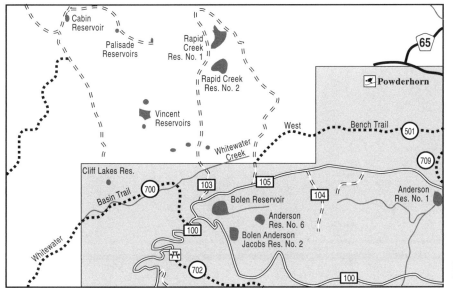

Map 1

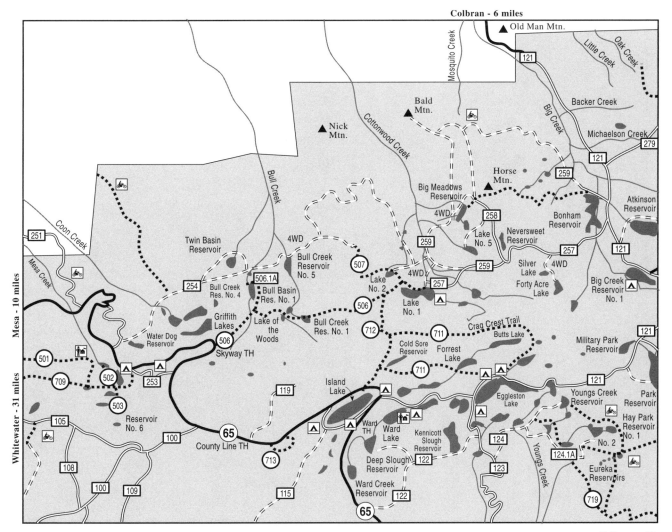

Map 2

Grand Mesa

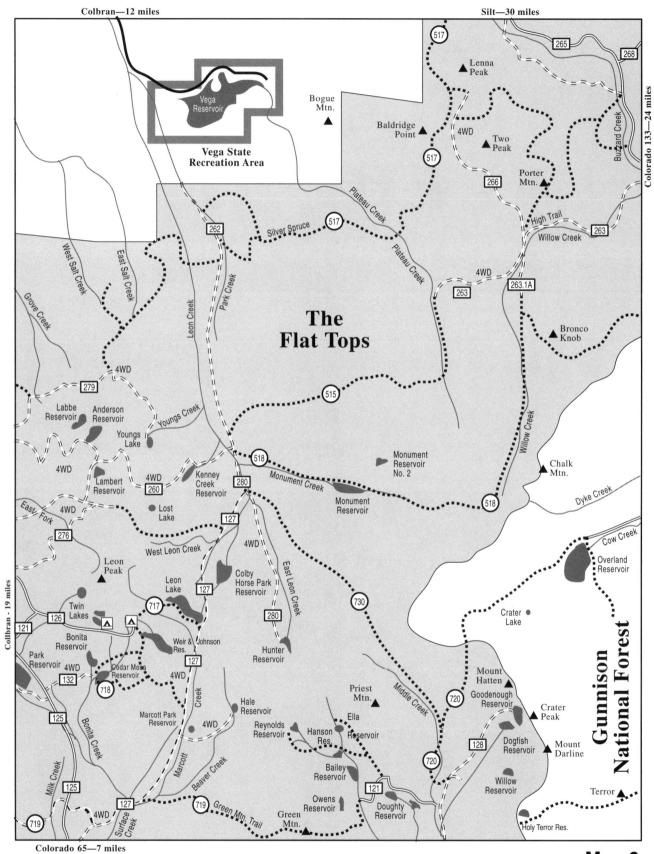

Map 3

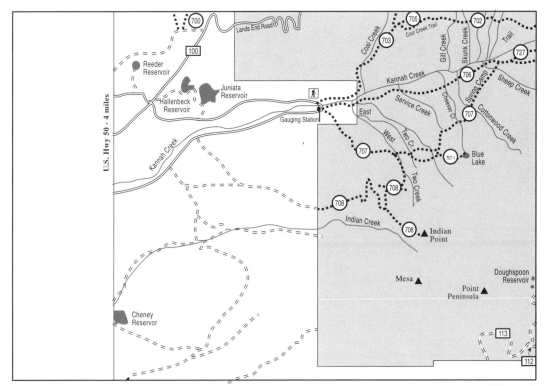

Map 4

Grand Mesa

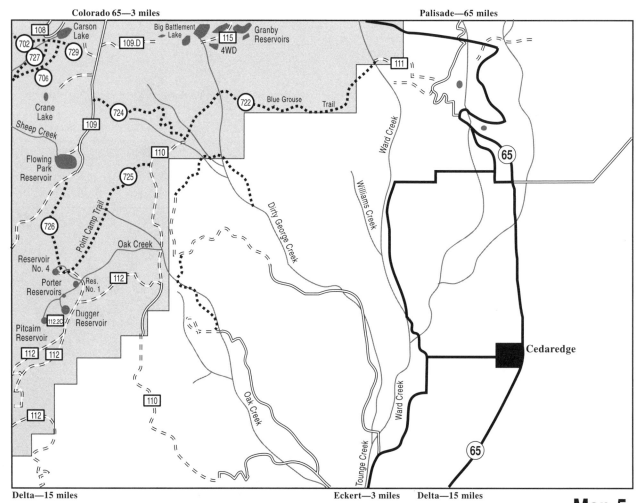

Map 5

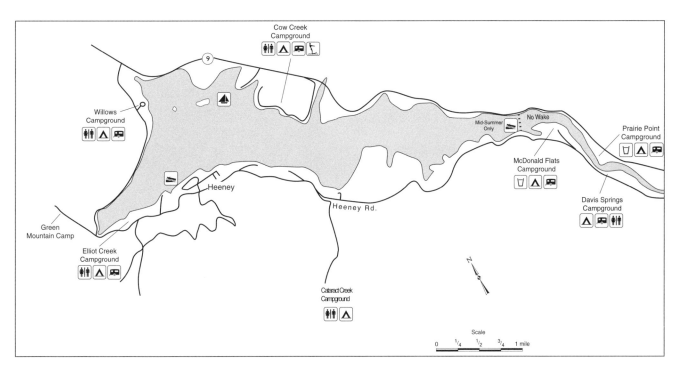

DIRECTIONS Take I-70 west to the Dillon exit. Travel north from Dillon on Colorado 9 for 16 miles to the south entrance (County Road 30). Follow County Road 30 around the west side of reservoir. Road parallels the reservoir continuing around to the north end and back to Colorado 9.

FEE Camping fee, except for Cow Creek Campground.

SIZE 2,125 acres.

ELEVATION 7,942 feet.

MAXIMUM DEPTH 195 feet.

FACILITIES Picnic tables, fire grates, water and trash removal for Prairie Point and McDonald Flats Campgrounds. No facilities for remaining four camping areas.

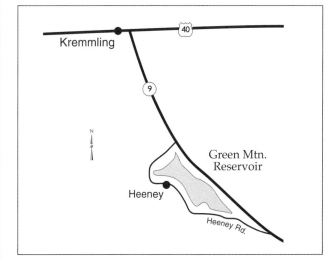

BOAT RAMP Concrete boat ramps are available at the McDonald Campground and at the Green Mountain Marina in the town of Heeney.

FISH Good for Kokanee salmon, fair to good for rainbow trout and fair for brown trout. Can also find cutthroat, lake trout, and sucker. For kokanee snagging season refer to current Division of Wildlife Regulations.

RECREATION Fishing, camping, personal watercraft, sailing, hiking.

CAMPING Six camping areas are available. Prairie Point and McDonald Flats Campgrounds are full service. The remaining four camping areas are open camping in a shadeless setting.
Elliott Creek—Toilets, trailers to 21 ft., no water.
McDonald Flats—13 sites, toilets, water, trailers to 21 ft.
Davis Springs—7 sites, toilets, no water.
Prairie Point—44 sites, tables, grates, water and toilets.
Cow Creek—Toilets, no defined spaces, no water.
Willows—Toilets, trailers to 32 ft., no water.

OTHER INFORMATION The Green Mountain Reservoir has a capacity of 152,000 acre-feet of water, 52,000 acre-feet of which is replacement of project diversions and 100,000 acre-feet of which is available for power purposes. Irrigation outlet capacity is 1,000 cubic feet per second, and power outlet capacity is 1,500 cubic feet per second. The spillway capacity is 25,000 cubic feet per second.

The reservoir is located on the Blue River, about 16 miles southeast of Kremmling, in a canyon north of the Gore Range of the Rockies. The reservoir extends up the Blue River for a distance of seven miles and has an area of 2,125 surface acres. The hills bordering the northeast side of the reservoir are, in general, more abrupt than those on

the southwest side and include a high outcropping of rock called Green Mountain, from which the dam takes its name. The area which was submerged by the reservoir consisted of hay and grazing land, with sage brush and scattered timber. The land surrounding the reservoir is virtually treeless.

CONTROLLING AGENCY White River National Forest.

INFORMATION Dillon Ranger District at (970) 468-5400.

MAP REFERENCES Benchmark's *Colorado Road & Recreation Atlas:* p. 73.

Gross Reservoir

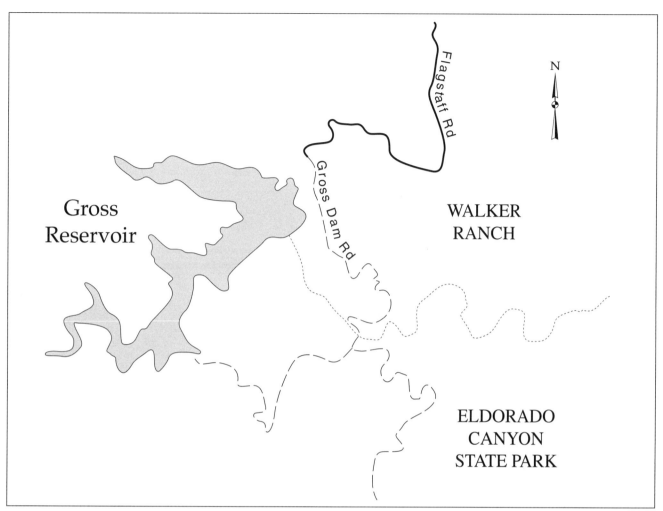

DIRECTIONS From Boulder, travel west on Baseline Road to Flagstaff Drive. Flagstaff Drive winds southwest about 7 miles to Gross Reservoir. It can also be reached through Coal Creek Canyon.

FEE No.

SIZE 440 surface acres.

ELEVATION 7,287 feet.

MAXIMUM DEPTH 230 feet.

FACILITIES Restrooms, fire rings, sheltered picnic area.

BOAT RAMP None. Motorized boating is not permitted from Memorial Day to the end of September.

RECREATION Fishing, picnicking and hiking.

SPECIAL RESTRICTIONS No water contact sports, such as swimming, waterskiing, wading, or scuba diving are permitted.

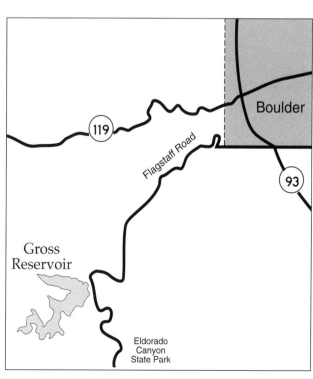

Gross Reservoir

CAMPING Camping is only allowed on US Forest Service property on west side of reservoir. There is limited access to these campsites, and high-clearance 4WD is required to access the west side that is managed by Roosevelt National Forest. The area has seasonal closures. Contact Boulder Ranger District for information (303) 541-2500. Fires allowed in fire pits on north side only.

FISH Gross Reservoir is open for fishing a half hour before sunrise to a half hour after sunset.Rainbow, brook, brown and lake trout. Also good for splake, kokanee salmon, and tiger muskie. Kokanee salmon snagging permitted September 1 to January 31. Floating devices prohibited.

CONTROLLING AGENCY Denver Water.

INFORMATION Division of Wildlife (303) 291-7227.

MAP REFERENCES Arapaho & Roosevelt National Forests map.
Benchmark's *Colorado Road & Recreation Atlas:* p. 62.

Groundhog Reservoir

DIRECTIONS   Travel 38 miles north of Dolores on Forest Road 526 to Forest Road 533, then 5 miles northeast to the reservoir.

FEE   No.

SIZE   688 acres.

ELEVATION   8,720 feet.

FACILITIES   Restrooms, drinking water, store.

BOAT RAMP   Public boat ramp. Wakeless boating.

FISH   Rainbow and cutthroat trout. Good fishing for rainbow trout, moderate pressure.

RECREATION   Fishing, boating.

CAMPING   There are 13 campsites and a store at the lake with food and a license agent.

CONTROLLING AGENCY   Division of Wildlife has fishing easement leased from Montezuma Valley Irrigation Company below high water line and at dam on boat ramp area.

INFORMATION   Division of Wildlife, Montrose Office (970) 252-6000.

MAP REFERENCES   San Juan National Forest map. Benchmark's *Colorado Road & Recreation Atlas:* p. 109.

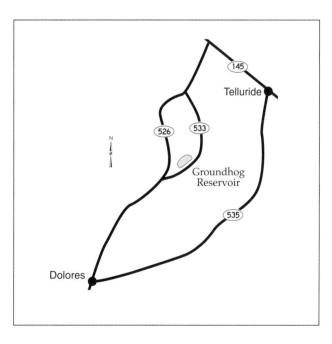

Harvey Gap State Park

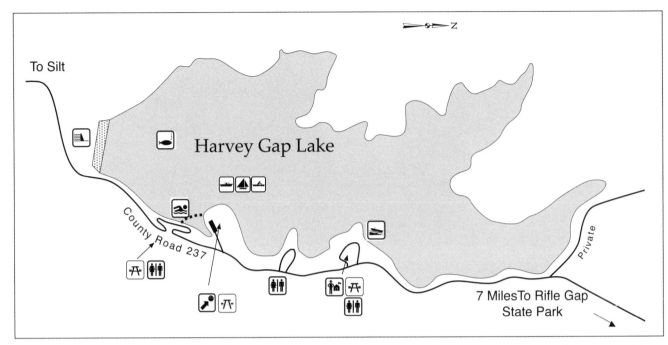

DIRECTIONS **Drive north** from Rifle 3 miles on Colorado 13 to Colorado 325, then travel northeast on Colorado 325. You will pass Rifle Gap Reservoir en route to Harvey Gap Reservoir State Park.

FEE Daily or annual pass.

SIZE 196 surface acres.

ELEVATION 6,500 feet.

FACILITIES Restrooms, picnic sites.

BOAT RAMP A concrete ramp is available on the south side of the lake. Boating is restricted to 20 horsepower motors or less.

FISH Harvey Gap Reservoir is one of the best fisheries in western Colorado. It features rainbow and brown trout, large and smallmouth bass, channel catfish, perch, crappie and northern pike. Rainbows take Power Bait, salmon eggs and nightcrawlers from shore or trolling Pop Gear with a nightcrawler. Browns are caught with nightcrawlers from shore in the early mornings or late evenings. Bass take topwater spinner baits or jigs and curly-trailed grubs. The best action has come from working jigs along the structure. Fish catfish with typical stink baits. Crappie take small white or yellow tube jigs suspended below a slip bobber and tipped with a piece of worm.

RECREATION Fishing, swimming, sailing, windsurfing, hunting, ice fishing, snowshoeing.

CAMPING No camping is available at the reservoir. It is a day-use area only. Harvey Gap Reservoir features cottonwood trees along the shoreline. It is in a typical high-plains

setting with the Grand Hogback mountain range providing a scenic backgound.

Camping is available nearby at Rifle Gap State Park (89 sites) and Rifle Falls State Park (13 sites). Reservations are advised throughout the summer. Campsites offer dump station, vault toilets, water, picnic tables, fire pits and grills.

CONTROLLING AGENCY Colorado State Parks.

INFORMATION Rifle office (970) 625-1607.

HANDICAPPED ACCESS Handicapped accessible restrooms, swimming, picnic tables, and fishing. Some facilities accessible with assistance.

MAP REFERENCES Benchmark's *Colorado Road & Recreation Atlas:* p. 70.

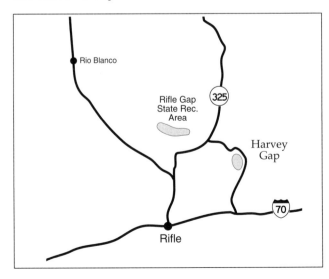

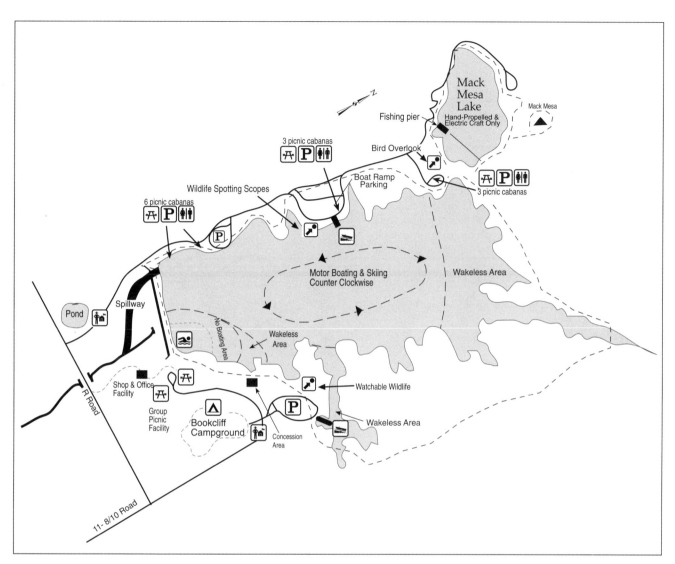

DIRECTIONS Travel west of Fruita on I-70 then take exit 15. Drive north on Colorado 139 approximately 8 miles to Q Road, then west 1.2 miles, and finally north one mile to reservoir.

FEE Daily or annual pass.

SIZE Highline—153 acres, Mack Mesa—15 acres.

ELEVATION 4,697 feet.

MAXIMUM DEPTH Highline—57 feet, Mack Mesa—30 feet.

FACILITIES Restrooms, dump station, water, picnic area, swim beach and food concessions. Group picnic area available. Lake View Rentals offers jet ski rentals, paddle boats, hydro bikes, and canoes.

BOAT RAMP Concrete boat ramps are located on both sides of Highline Reservoir. Small boat ramp at Mack Mesa for use with hand or electric-powered boats.

FISH Mack Mesa Lake is noted for its fine early season trout fishing. Mack Mesa Lake provides good fishing and solitude away from motorized boats. Only hand or electric powered boats are allowed. Highline Lake has good warm-water fishing, especially for catfish and crappie, as well as largemouth bass and black bullhead.

SPECIAL RESTRICTIONS There is a 40 boat capacity restriction on weekends. If the lake is at capacity wait at west entrance for a boat to come off the lake.

RECREATION Fishing, swimming, waterskiing, boating, picnicking, hunting, hiking, birding, biking and OHV use.

CAMPING Highline State Park offers 31 grassy campsites that can accommodate both tents and RVs. Showers and laundry facilities are available. A holding tank dump station is located in the campground. There are no electrical hookups.

HANDICAPPED ACCESS A fishing jetty with wheelchair stops is located on the southwest side of Mack

Highline Lake State Park

Mesa Lake and is designed specifically to accommodate persons with disabilities. The swim area is located 50-75 yards from the parking lot and is accessible with assistance. The beach house, toilet facilities and group shelter are all accessible. One handicap-accessible campsite is available. Campsites are level with graveled pads and lawn, and a concrete walkway is accessible along the lake shore.

OTHER INFORMATION More than 150 species of birds have been observed at Highline Lake State Park, and migrating ducks and geese winter at Highline Lake. Birds such as the great blue heron, white pelican, snowy egret, whooping crane, golden eagle and bald eagle are seen in the park. Many small animals make their home at Highline, and natural areas within the park allow visitors and school groups to observe animals in their natural habitats.

CONTROLLING AGENCY Colorado State Parks.

INFORMATION Lorna Office (970) 858-7208.

MAP REFERENCES Benchmark's *Colorado Road & Recreation Atlas:* p. 82.

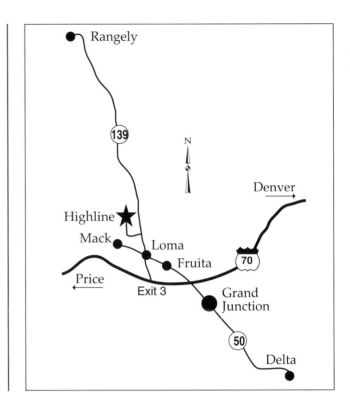

Horsetooth Reservoir

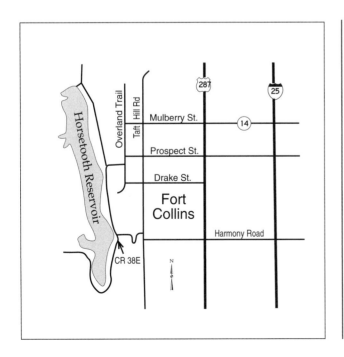

DIRECTIONS From Denver drive north on I-25 to Fort Collins and take the Harmony-Timnath exit. Drive west on Harmony Road to a T-intersection. Turn right (north) onto Taft Hill Road and travel about one mile. Turn left (west) onto County Road 38E, which will take you to the reservoir.

FEE Daily or annual pass.

SIZE 1,899 acres.

ELEVATION 5,430 feet.

MAXIMUM DEPTH 180 feet.

FACILITIES This is a complete service recreation area that includes a marina and food store.

BOAT RAMP One ramp is available at Satanka Bay on the north end, two at South Bay Landing on the south end and one at Inlet Bay Marina on the southwest side. All ramps are concrete.

Horsetooth Reservoir

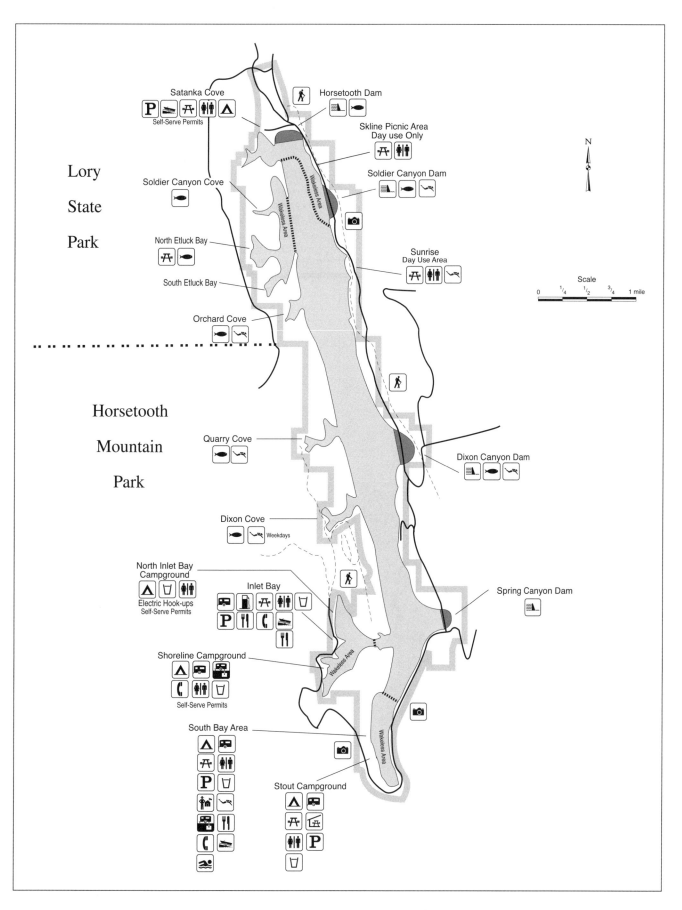

Horsetooth Reservoir

FISH Largemouth and smallmouth bass, bluegill, crappie, carp, splake, wipers, rainbow trout, walleye and perch. Fair to good for trout, either shore fishing with salmon eggs, worms or Power Bait and trolling flashers tipped with a nightcrawler. Fishing good for smallmouth bass and occasional largemouth using typical lead-head jigs and curly tails as well as crankbaits. Most bass are 2-3 pounds, with some up to 4 pounds. Bass under 15 inches must be released. Walleye fishing can be great during the early spring months trolling bottom bouncers with minnows or nightcrawlers and assorted crankbaits. Trophy walleye up to 16 pounds, with many in the 10-12 pound range, are caught every year. Walleye restricted to an 18-inch minimum. Perch fishing is fair to good on nightcrawlers and a slip bobber in most of the coves.

Routine sampling has shown that some fish from this water exceed the mercury action level of 0.5 parts per million set by the Colorado Department of Public Health and Environment. For more information, visit *http://www.cdphe.state.co.us/wg/monitoring/monitoring.html.*

RECREATION All water recreation sports, scuba diving, fishing and camping.

CAMPING There are 180 designated campsites between Satanka Bay, Inlet Bay and South Bay campgrounds. All campgrounds feature vault toilets, picnic tables and fire grates. Open camping also is allowed where access allows along shorelines of reservoir. Dump stations are located at South Bay and Inlet Bay. Cabins are available to rent.

CONTROLLING AGENCY Larimer County Parks and Open Lands Dept.

INFORMATION Park office (970) 679-4570.

OTHER INFORMATION Horsetooth Reservoir, located just west of Fort Collins and adjacent to Lory State Park, was constructed as part of the Big Thompson Water Project. The reservoir is 6.5 miles long and covers 1,875 surface acres. It was formed by the construction of four large earth-filled dams. Horsetooth, Spring Canyon, Dixon Canyon and Soldier Canyon Dams, each approximately 210 feet high, are built across openings in the long hogback forming the eastern side of the reservoir. Horsetooth Dam is constructed across the north end of the foothill "glade" formed by the uplift ridges.

The reservoir covers what once was an old quarry town named Stout. Stout was constructed by the Greeley, Salt Lake and Pacific Railroad, under the charter and backing of the Union Pacific Railroad. The area had an almost limitless supply of high grade fine-grained sandstone; the vari-

ety of color and quality of the sandstone made it popular for use as building material in the late 1800s. Boats using sonar for fish finding may locate some old structures left standing when Horsetooth Reservoir was filled. Other traces of quarry activity may be found on the west side of the reservoir.

The scenic setting and the variety of recreation activities makes Horsetooth Reservoir one of the top recreation areas on the front range.

HANDICAPPED ACCESS Handicapped accessible restrooms located at South Bay, Inlet Bay and all campgrounds.

MAP REFERENCES Benchmark's *Colorado Road & Recreation Atlas:* p. 49.

Jackson Lake State Park

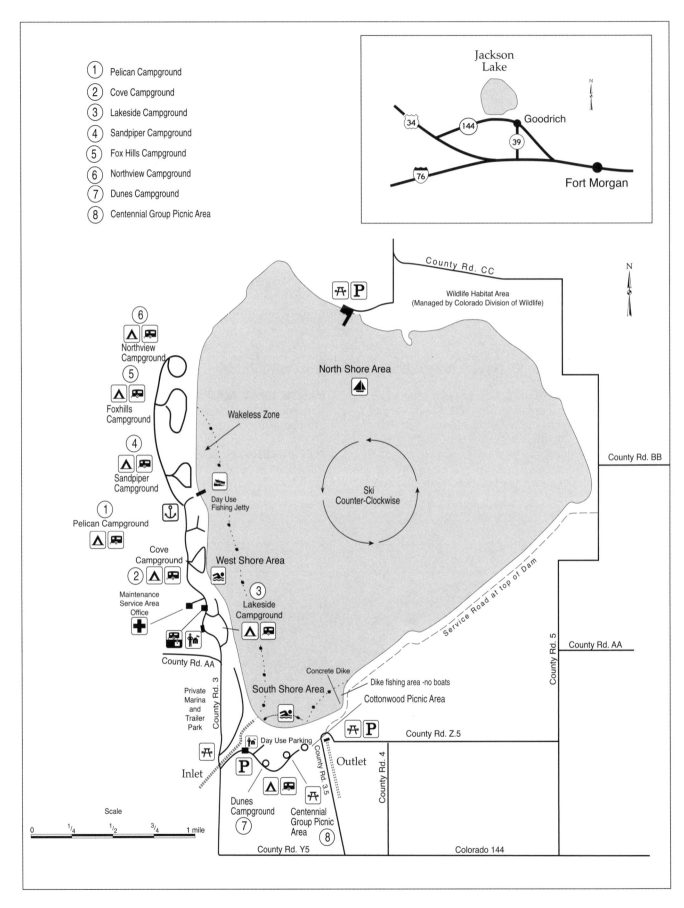

1. Pelican Campground
2. Cove Campground
3. Lakeside Campground
4. Sandpiper Campground
5. Fox Hills Campground
6. Northview Campground
7. Dunes Campground
8. Centennial Group Picnic Area

Jackson Lake

34
144
39
Goodrich
76
Fort Morgan
N

County Rd. CC

Wildlife Habitat Area
(Managed by Colorado Division of Wildlife)

N

North Shore Area

6
Northview Campground

5
Foxhills Campground

Wakeless Zone

Ski Counter-Clockwise

County Rd. BB

4
Sandpiper Campground

1
Pelican Campground

Day Use Fishing Jetty

West Shore Area

Cove Campground

2

3
Lakeside Campground

Maintenance Service Area Office

Service Road at top of Dam

County Rd. 5

County Rd. AA

County Rd. AA

County Rd. 3

Concrete Dike

South Shore Area

Dike fishing area -no boats

Cottonwood Picnic Area

Private Marina and Trailer Park

Day Use Parking

County Rd. Z.5

County Rd. 4

Outlet

Inlet

Dunes Campground
7

County Rd. 3.5

Centennial Group Picnic Area
8

County Rd. Y5

Colorado 144

Scale

0   1/4   1/2   3/4   1 mile

DIRECTIONS From Denver travel northeast on I-76 to the US 34 interchange. County Road 39 north 7.2 miles to Goodrich. Travel west on Y5 (follow the paved road) for 2 miles to the Jackson Lake State Recreation Area.

FEE Daily or annual parks pass.

SIZE 2,700 acres.

ELEVATION 4,438 feet.

MAXIMUM DEPTH 30 feet.

FACILITIES Marina, boat and jet ski rentals, visitor center, store, showers, toilets, drinking water, picnic tables, laundromat and fire pits.

BOAT RAMP One concrete boat ramp on the west shore with four lanes and a courtesy dock located near the marina. Bait, food, gas, camping supplies, boating and fishing supplies are available at the marina.

RECREATION Fishing, OHV riding, biking, hiking, hunting, waterskiing, personal watercraft, sailing, swimming.

CAMPING There are 260 campsites available and can accommodate campers, trailers or tents. Campground settings vary from sweeping sandy beaches to stands of stately cottonwoods. There are electrical hookups, showers, toilets and drinking water. A trailer dump station is located on the west side of Jackson Lake. Reservations are recommended for any weekend. Jackson Lake State Park offers two sandy swim beach areas.

FISH Trout, bluegill, saugeye, walleye, catfish, perch, crappie and wiper. Fishing is fair to good for all species. Trout are caught with Power Bait and salmon eggs from shore, trolling with Pop Gear and a nightcrawler. Walleye and wiper are caught trolling with a Bottom Bouncer or nightcrawler harness. Catfish are caught with cutbait, chicken liver and assorted stink baits. Perch and crappie are caught jigging small white or chartreuse jigs.

HANDICAPPED ACCESS Handicapped accessible showers, restrooms, hunting and fishing. One reserved site at Lakeside Campground, restrooms and picnic tables throughout park. Fishing from boat ramp. Some facilities accessible with assistance.

CONTROLLING AGENCY Colorado Division of Parks.

INFORMATION Orchard Office (970) 645-2551.

OTHER INFORMATION The reservoir was built in 1902 and incorporates an existing lake. Its 2,700 surface acres provide irrigation for the thousands of acres of farmland to the south and east. The area surrounding the lake became part of Colorado's system of state parks and state recreation areas in 1965.

Fishermen come here for walleye, wiper, catfish, perch, crappie and trout; but fishing is restricted to designated areas during migratory waterfowl season.

Jackson Lake's sandy beaches and gradually sloping lake bottom make it ideal for swimming.

Some 200 acres on the north side of the lake are maintained in a primitive state as wildlife habitat—a walk-in, no camping area. It attracts visitors who wish to observe or photograph the abundant wildlife found in the area including ducks, geese, pelicans, pheasant, bald and golden eagles, rabbits, coyote and much more.

SPECIAL RESTRICTIONS

Place litter in the receptacles provided. Keep your vehicle on designated roads or parking spaces. Keep your pets on a leash six feet long or less. Camp only in designated camping sites and bring no more than the permitted number of vehicles to each site. Report any problems to a park ranger.

All of Jackson Lake areas are open to waterskiing. Boats towing skiers must stay 150 feet from the swim area, moored vessels, and shore fishermen. Skiing is counter clockwise. No boats are allowed in or near the swim beaches and the lake is closed to boating from November through the waterfowl season.

Swimming is restricted to two designated swimming areas and is at the swimmer's own risk. Dogs, glass containers, fires and fishing are not permitted on the swim beaches.

Hunting is permitted in designated areas only.

MAP REFERENCES Free park map.
Benchmark's *Colorado Road & Recreation Atlas:* p. 65.

James M. Robb—Colorado River State Park

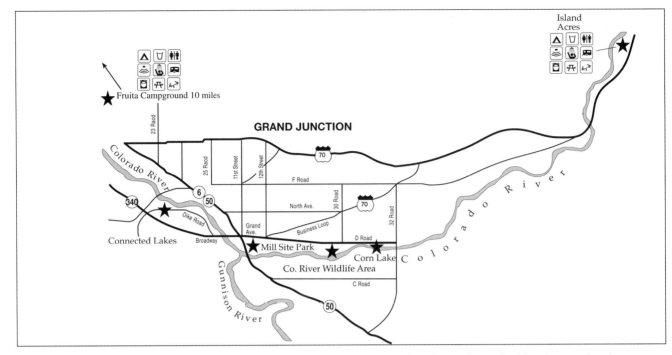

DIRECTIONS The state park runs from Grand Junction along the Colorado River east to Island Acres.

FEE Daily or annual pass. Camping fee. There is no fee for hikers and bicyclists.

ELEVATION 5,000 feet.

FACILITIES Restrooms, water.

BOAT RAMP Corn Lake, Connected Lakes, and Fruita are launching sites for boaters and rafters on the Colorado River.

FISH Cold water and warm water fishing. Islands Acres, Corn Lake, and Fruita are stocked with trout in the spring and fall.

RECREATION Fishing, OHV riding, biking, hiking, boating, rafting, swimming.

CAMPING There are 74 sites at Island Acres, and 63 sites at Fruita most with full hookups. RV dump station and tent camping are available at both sites. At Fruita a service building offers campers flush toilets, laundry and playground.

CONTROLLING AGENCY Colorado State Parks.

INFORMATION Clifton Office (970) 858-9188.

CORN LAKE Day use area located at 32 Road and the Colorado River. Corn Lake provides a launching site for boaters and rafters to the Colorado River. Trails provide access to the Colorado River and Corn Lake for fishing and are used by hikers and bicyclists. The park also offers picnic sites and restrooms accessible to the physically challenged. There is no charge for bicycle or pedestrian access to the park.

CONNECTED LAKES Day use area, accessed by travelling north on Dike road off Colorado 340. Connected Lakes provides a network of trails traversing a series of reclaimed gravel pits giving visitors a wide variety of recreational opportunities including fishing, picnicking, hiking and bird watching. There is no charge for bicycle or pedestrian access to the park.

ISLAND ACRES Located at exit 47 on I-70 in DeBeque Canyon. The park is open year round for camping and day-use activities with the day-use closing at 10:00 p.m. The park is a convenient and attractive place to fish, swim, camp, picnic and/or hike along the Colorado River or near any of the lakes in the park.

COLORADO RIVER WILDLIFE AREA Located one mile west of Corn Lake on D Road. Hiking, wildlife observation and environmental education activities available. Fishing access is on the south and west sides of the lake only.

OTHER INFORMATION Enjoy the many recreational activities available in Fruita township including swimming, hiking, and scenic views of Colorado National Monument.

HANDICAPPED ACCESS The Corn Lake and Connected Lakes sections provide accessible picnic areas, parking areas, and restrooms. The Island Acres and Fruita sections provide accessible parking areas, restrooms and campsites. Corn Lake and Island Acres provide a wheelchair accessible concrete fishing pier, and Connected Lakes has four sheltered, accessible fishing sites.

MAP REFERENCES Benchmark's *Colorado Road & Recreation Atlas:* p. 83.

Jefferson Lake

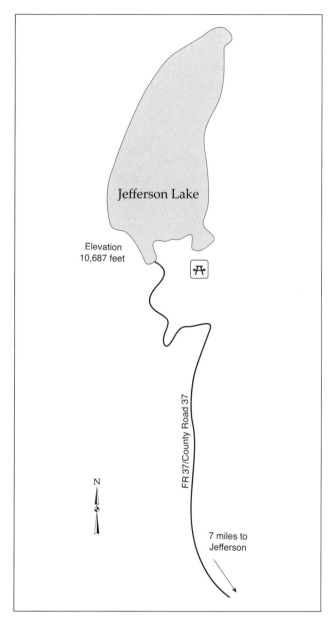

DIRECTIONS From Denver, travel south on US 285 over Kenosha Pass toward the town of Jefferson. From Jefferson, head west on County Road 35. Turn right on County Road 37 and head north toward the lake, about 5 miles.

FEE Daily or annual pass. Camping fee.

SIZE 175 surface acres.

ELEVATION 10,687 feet.

FACILITIES Restrooms, drinking water, picnic tables.

BOAT RAMP A concrete boat ramp is located on the southeast end of the lake. There are no motor size restrictions but boats are restricted to no white water wake.

FISH Stocked kokanee salmon, rainbow, brook, brown, lake, and cutthroat trout. Rainbow; can be caught from shore with green Power Bait, Fireballs and worms. Good results also can be ontained from trolling flashers, Pop Gear, trailing a nightcrawler or lure. Brook; various fly patterns work with a casting bubble to rising fish in the late evening. Lake; trolling Rapala or Kastmaster lures at a depth of 30-50 feet or jigging Gitzits tipped with sucker meat along the bottom structure.

RECREATION Fishing, camping, hiking.

CAMPING Several campgrounds are available along the US Forest Service access road. Camping is restricted to designated campgrounds, and no camping is allowed along the shore. Camping is first-come, first-served. Reservations are not accepted. Three fee-area campgrounds, offering a total of 64 campsites, are available. All areas feature restrooms, drinking water, picnic tables and fire grates. Jefferson Lake features a tree-covered shoreline and a high-country setting along the Continental Divide. It is one of the most scenic lakes in the South Park region.

CONTROLLING AGENCY Pike & San Isabel National Forests.

INFORMATION South Park Ranger District (719) 836-2031.

MAP REFERENCES Pike National Forest map. Benchmark's *Colorado Road & Recreation Atlas:* p. 73.

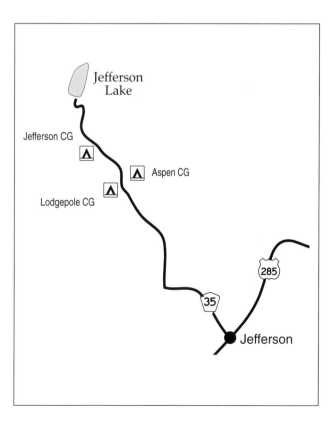

John Martin Reservoir State Park

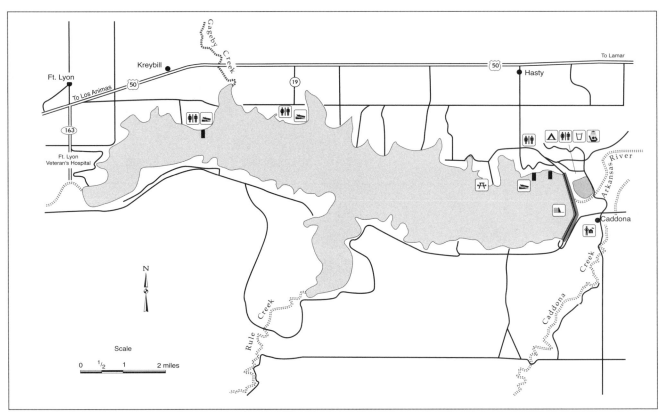

DIRECTIONS From Las Animas travel 16 miles east on US 50 to Hasty then go two miles south to the reservoir.

FEE Annual or daily pass.

SIZE Varies greatly from 1,000 to over 11,000 acres.

ELEVATION 3,800 feet.

MAXIMUM DEPTH 60 feet.

FACILITIES Restrooms, camper service building, fishing piers, swim beach, picnic sites, trails, visitor center, group picnic and camping facilities.

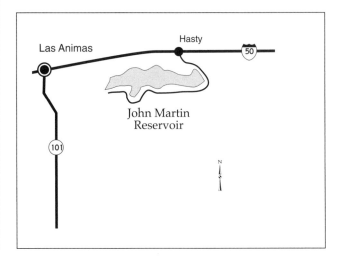

John Martin Reservoir State Park

BOAT RAMP Two concrete ramps. The east ramp is shallow, the west ramp is steep.

FISH Bass, catfish, walleye, perch, wipers, bluegill, crappie and saugeye. A large, plains reservoir that can produce large fish from the shore or trolling; wind is fairly common.

RECREATION Boating, waterskiing, sail boarding, personal watercraft, fishing.

CAMPING There are 213 campsites in two separate campgrounds. Lake Hasty Campground is open year-round and has electrical hookups at each of the 109 sites. It is a full service campground with showers, laundry facilities, a dump station, and a fish cleaning station. Point Campground on the north shore offers basic camping with vault toilets, fire grills, and picnic tables.

CONTROLLING AGENCY Colorado State Parks.

INFORMATION Hasty Office (719) 829-1801.

HANDICAPPED ACCESS Handicapped Accessible Facilities. Restrooms, showers and the fishing pier on Lake Hasty near campground.

OTHER INFORMATION The dam which created John Martin Reservoir was built between 1939 and 1948 as an irrigation and flood control project by the US Army Corps of Engineers. Colorado State Parks has managed the Lake Hasty area below the dam and the surface of the reservoir since 2001 through a lease agreement with the Army Corps of Engineers.

The sand and gravel shores of John Martin Reservoir are among the few remaining nesting areas in Colorado for the threatened Piping Plover and the endangered Interior Least Tern. In an effort to conserve these species, portions of the shoreline and the reservoir surface are temporarily closed to all public access during the resting and brooding season. Some campsites are closed from November 1 to March 31 to protect the bald eagles.

MAP REFERENCES Benchmark's *Colorado Road & Recreation Atlas:* p. 119.

Lake Avery

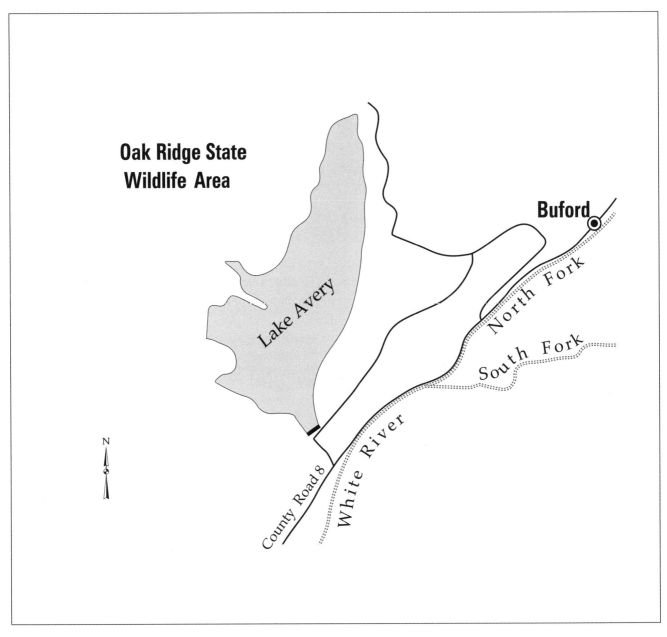

DIRECTIONS Travel east from Meeker on Colorado 13 to County Road 8. Continue east on County Road 8 about 17 miles. The access road to the lake is located on the north side of the road, just before the town of Buford.

FEE No.

SIZE 245 surface acres when full.

ELEVATION 6,985 feet.

MAXIMUM DEPTH 76 feet.

FACILITIES Fire rings, pit toilets, picnic tables, and a dump station are available.

BOAT RAMP A concrete ramp is available, but its usability depends on the lake's water level. The level usually is good through June.

RESTRICTIONS Wakeless boating only.

FISH Rainbow trout, some brook and cutthroat trout. Fishing is rated good for 10- to 15-inch rainbows, with some taken up to 5.5 pounds. Early fishing is excellent as open water becomes available, usually from mid-April to mid-June. Summer weed growth hampers bank fishing. Trolling from a boat gets the best results.

RECREATION Fishing, boating, hiking, hunting, horseback riding, and wildlife viewing.

Lake Avery

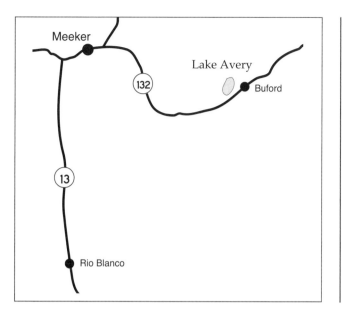

CAMPING This property is part of the Division of Wildlife's Oak Ridge State Wildlife Area. A primitive campground offers open camping. The campground generally is used mostly in May and June. The area around the lake is mostly open, offering surrounding mountain scenery of White River National Forest.

CONTROLLING AGENCY Division of Wildlife.

INFORMATION Grand Junction Office (970) 255-6100.

MAP REFERENCES White River National Forest map. Benchmark's *Colorado Road & Recreation Atlas:* p. 58.

La Jara Reservoir

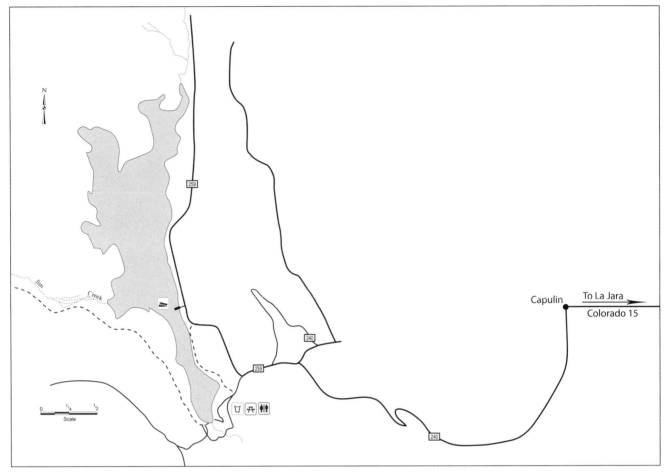

DIRECTIONS   From La Jara travel north on US 285 to the intersection with Colordo 15. Travel west on Colorado 15 for about 9 miles to Capulin. In Capulin, turn right onto County Road 8 and follow the road to the intersection with Forest Road 240. Follow Forest Road 240 for approximately 6 miles to the intersection with Forest Road 259. Follow Forest Road 259 to the northwest for about 4 miles to the reservoir.

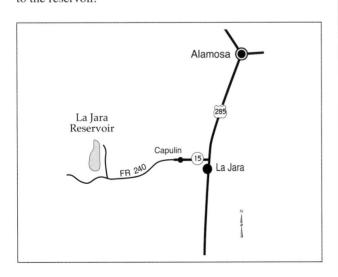

FEE No.

SIZE 1,375 acres.

ELEVATION 9,698 feet.

FACILITIES Restrooms, drinking water.

BOAT RAMP Ramp is located on the east side of reservoir.

FISH Brook trout averaging 8 inches. Shallow weedy reservoir.

RECREATION Fishing, hunting, boating, birding, wildlife viewing.

CAMPING Dispersed camping.

CONTROLLING AGENCY Colorado Division of Wildlife.

INFORMATION Monte Vista Office (719) 587-6900.

MAP REFERENCES Benchmark's *Colorado Road & Recreation Atlas:* p. 124.

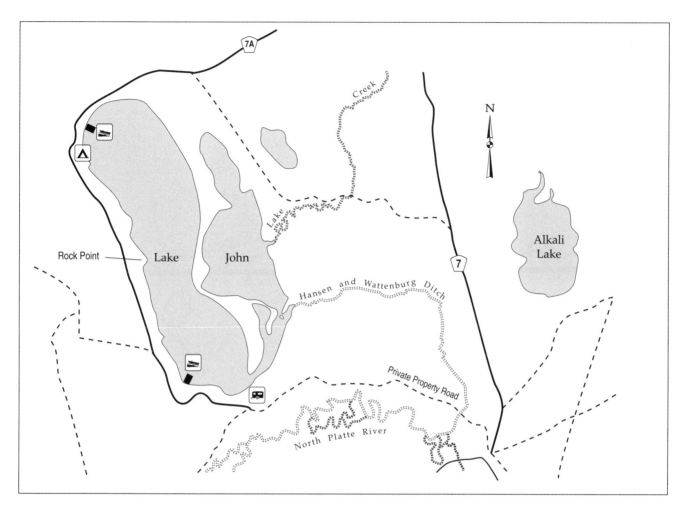

DIRECTIONS From Walden travel 0.5 miles west on Colorado 14 to County Road 12, then go 8 miles west to County Road 7, then a farther 7 miles north to lake.

FEE Habitat Stamp required.

SIZE 656 acres.

ELEVATION 8,048 feet.

FACILITIES Restrooms, dump station.

BOAT RAMP Located at north and south end of lake.

FISH Rainbow, brown, cutthroat trout, and sucker. Rated excellent. Check regulations for bag limits.

RECREATION Fishing. All boats allowed, but no waterskiing or sail boarding.

CAMPING Lake John RV Park and Cabins. Full service restaurant on weekends. General store, ice shanty rentals, and fishing holes drilled. There are 30 RV sites with full hookups. Call (970) 723-3226 for reservations. Dispersed camping at state areas (14 day maximum).

CONTROLLING AGENCY Colorado Division of Wildlife.

INFORMATION Ft. Collins Office (970) 472-4300.

MAP REFERENCES Benchmark's *Colorado Road & Recreation Atlas:* p. 46.

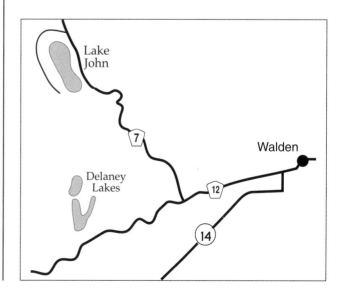

Lake Pueblo State Park

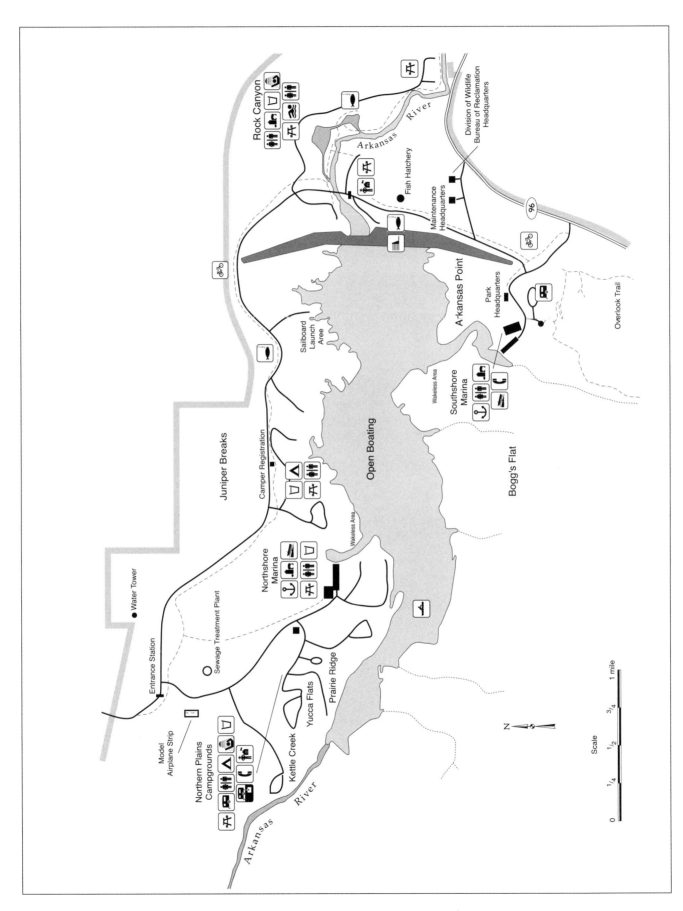

DIRECTIONS The park has two entrances. To get to either from I-25, begin at the intersection with US 50 in Pueblo.

For the south park entrance: Turn west on US 50 (I-25 exit 101). Drive 4 miles to Pueblo Boulevard. Turn south and go 4 miles to Thatcher Boulevard (Colorado 96). Turn west and go 4 miles to the park entrance.

For the north park entrance: Turn west on US 50 (I-25 exit 101). Drive 7 miles to McCulloch Boulevard. Turn south and go 4 miles to Nichols Road. Turn south and go 1 mile to the north park entrance.

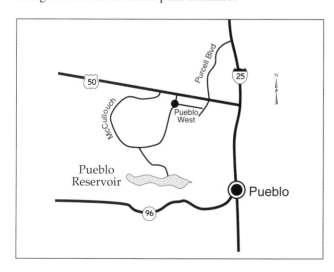

FEE Daily or annual pass.

SIZE 4,646 acres when full.

ELEVATION 4,880 feet.

MAXIMUM DEPTH 135 feet.

FACILITIES Full service state park with dump stations and marinas.

BOAT RAMP Two ramps are located at the north shore marina and the south shore marina.

RECREATION All water sports, bike trails, interpretive trail, hunting, horseback riding, picnicking, swimming.

CAMPING Lake Pueblo's 401 campsites accommodate tents, pickup campers, motor homes and trailers. Arkansas Point, Prairie Ridge and Yucca Flats campgrounds feature modern toilets, showers, laundry facilities and electricity. Juniper Breaks and Kettle Creek offer primitive facilities with vault restrooms but no running water. Group campgrounds are also available. Dump stations are located in the campgrounds. All campers must register at campground entrance before taking a site. Reservations are suggested April 1 through September 30.

FISH Fishing is the main attraction at Pueblo. Many anglers make this their home base. Both bank and boat fishing are productive. Rainbow trout dominate the fishing action here. Black crappie run a close second, followed by walleyes, small mouth bass, bluegills, channel catfish, tiger muskie, carp, cutthroat trout, sunfish, large-mouth bass and yellow perch and wipers. Ice fishing is popular in winter. People who wish to fish at night may stay on the water if their boat is equipped with running lights and a proper anchor.

OTHER INFORMATION Pueblo Reservoir in south central Colorado was developed as part of the Fryingpan/Arkansas Project. The major feature of the project is massive Pueblo Dam (210-feet-high and 1-mile-long), planned and built by the US Water and Power Resources Service. The project diverts water from the Colorado River Basin on the western slope, providing water for drinking and irrigation, hydroelectric power, wildlife and recreation. It also has considerable flood control value.

Since 1975, the Colorado Division of Parks and Outdoor Recreation has managed a 12,000 acre area surrounding the reservoir under lease from the federal government. An additional 6,000 acres are managed by the Colorado Division of Wildlife. Pueblo State Recreation Area, which includes the reservoir, appears to stretch endlessly eastward, while the Sangre de Cristo and Greenhorn mountain rages form an alpine backdrop to the west.

Limestone cliffs and flattop buttes rim the reservoir's irregular 60 mile shoreline. Vegetation includes prickly pear and cactus, along with cottonwoods and willows. Red-tailed hawks inhabit the area.

The campground is patrolled for safety and to ensure quiet after 10 PM. There are picnic sites throughout the park and two group picnic shelters near the north shore marina that may be reserved. There is a second marina on the south shore. Both marinas offer boat slips, boat rentals, food and fuel.

There is plenty of space for open boating and waterskiing. Floating debris and changing shorelines are a constant hazard. Boaters should keep a sharp watch for shallow areas and floating or partially submerged objects. Swimming is only permitted at the beach below the dam in Rock Canyon where lifeguards are on duty.

A wildlife management area at the eastern end of the reservoir provides cover and food for waterfowl and game. Open to hunters only during legal seasons, the area is administered by the Division of Wildlife.

HANDICAPPED ACCESS Handicapped accessible campsites in each campground near accessible restrooms. There is a fishing pond in Rock Canyon directly below the dam designed especially to accommodate disabled persons. It has a pier and a paved trail that surrounds the pond. Accessible visitors center, showers, restrooms, swimming, picnic tables, trails, camping and fishing. Some facilities accessible with assistance.

CONTROLLING AGENCY Colorado State Parks.

INFORMATION Pueblo Office (719) 561-9320.

MAP REFERENCES Benchmark's *Colorado Road & Recreation Atlas:* p. 101.

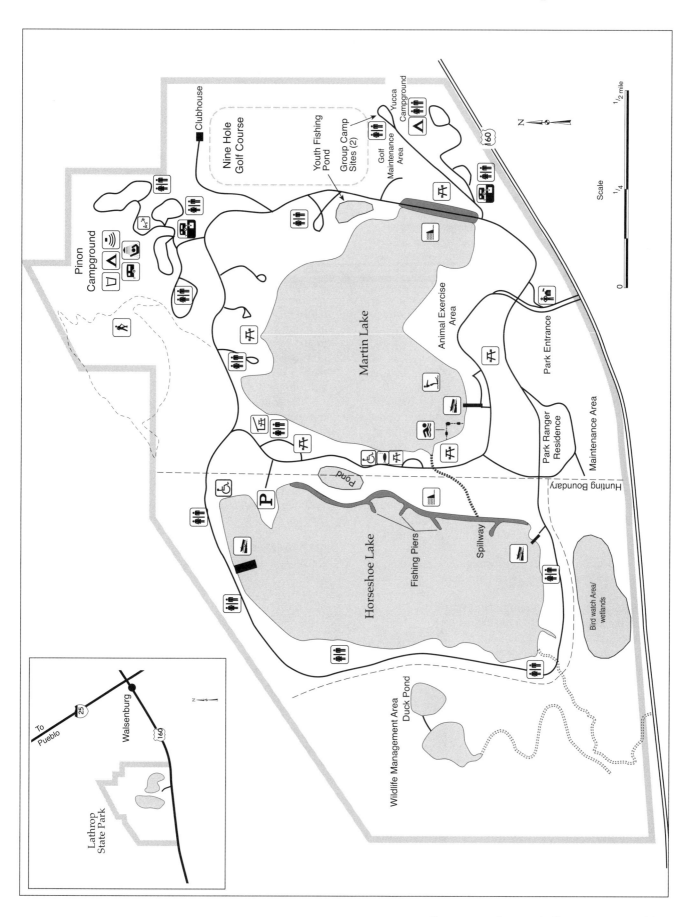

Clubhouse

Nine Hole Golf Course

Youth Fishing Pond

Group Camp Sites (2)

Yucca Campground

Golf Maintenance Area

Pinon Campground

Martin Lake

Animal Exercise Area

Park Entrance

Park Ranger Residence

Maintenance Area

Hunting Boundary

Pond

Horseshoe Lake

Fishing Piers

Spillway

Bird watch Area/ wetlands

P

Wildlife Management Area Duck Pond

160

N

Scale

1/2 mile

1/4

0

To Pueblo

25

Walsenburg

160

Lathrop State Park

N

Lathrop State Park

DIRECTIONS Take I-25 south from Denver to Walsenburg. Go three miles west of Walsenburg on US 160. The state park is on the north side of the highway.

FEE Daily or annual park pass.

SIZE Martin Lake—180 surface acres; Horseshoe Lake—150 surface acres.

ELEVATION 6,410 feet.

MAXIMUM DEPTH Horseshoe—25 feet; Martin—27 feet.

FACILITIES Dump stations and water pumps are available throughout both campgrounds. Additionally, there are restrooms, a visitor center, picnic areas, a swim beach and a 9 hole golf course, .

BOAT RAMP Concrete ramps on the north and south side of Horseshoe. Ramp located on the south side of Martin Lake. Horseshoe Lake is wakeless boating only.

FISH Rainbow trout, channel catfish, pike, wiper, bass, walleye, bluegill, crappie and tiger muskies. Tiger muskies in Horseshoe must be 30 inches or longer, limit one fish. Rainbow trout are caught on salmon eggs, Power Bait and nightcrawlers from the shore or by trolling pop gear tipped with a nightcrawler. Best results are in the early morning or late evening. Catfish are caught with stink baits, chicken liver and nightcrawlers from the shore at night. Walleye fishing is best on harnessed nightcrawlers or bottom bouncers. Bass are caught on spinner baits, top water baits and crank baits. Bluegill and crappie are caught on crappie jigs, worms below a bobber and flies. Smallmouth and largemouth bass possessed must be 15 inches or longer.

RECREATION Boating, sailing, swimming, personal watercraft, golf, water sports, hiking and bird watching.

CAMPING There are two campgrounds - Yucca and Pinion campgrounds. Lathrop State Park has 103 campsites and a group camping area. Yucca is non-electric and has vault toilets, picnic tables, fire pits, and a dump station. Pinon has electrical hookups, showers, restrooms, and a dump station. The campgrounds will accommodate motor homes, trailers and tents. Shower and laundry facilities.

CONTROLLING AGENCY Colorado State Parks.

INFORMATION Walsenburg Office (719) 738-2376.

SPECIAL RESTRICTIONS Pets must be on a leash not longer than 6 feet and clean up after them or leave them at

home. No pets on beach. Keep vehicles on roads or parking lot. Do not cut trees or gather dead wood for fires. Camp only in designated areas. Hunting is allowed only on the west end of the park, small game and water fowl only, short guns only. Hunting is restricted to the period of time after Labor Day and before Memorial Day.

OTHER INFORMATION Lathrop State Park offers its numerous visitors a place to relax and camp in the pinion-juniper and high plains grassland setting typical of southeastern Colorado. It is an ideal departure point for persons wishing to see the region's many natural and historical features. Within the park are facilities for picnicking, fishing, hiking and exciting sail boating.

Seeming to welcome and protect visitors, the Spanish Peaks (the west peak is 13,610 feet high and the east peak 12,669 feet high) rise in the distance like sentinels over an area rich in history, geology, culture and legends.

The early Indians named the peaks "Huajatolla" (Wa-ha-toya) and gave this religious description: "Two breasts as round as women's and all living things on earth, mankind, beasts and plants derive their sustenance from that source. The clouds are born there and without clouds there is no rain, and where no rain falls we have no food and without food we must perish all."

The state leased this 1,434-acre park in 1962 with the assistance of many local citizens. It is named after Harold W. Lathrop, the first director of the Colorado Division of State Parks which administers the area.

HANDICAPPED ACCESS Handicapped accessible visitor center, restrooms, swimming, picnic tables, camping and fishing. Some facilities accessible with assistance.

MAP REFERENCES Benchmark's *Colorado Road & Recreation Atlas:* p. 115.

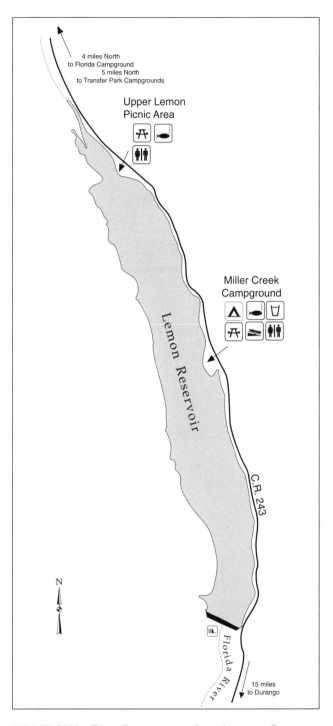

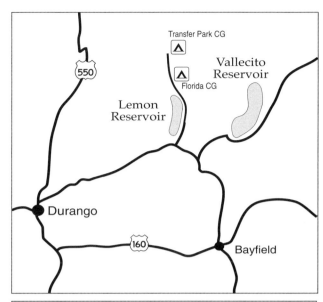

DIRECTIONS From Durango travel northeast on County Road 240 (also known as Florida Road) to County Road 243. Travel north to Lemon Reservoir.

FEE Camping fees.

SIZE 622 surface acres.

ELEVATION 8,148 feet.

FACILITIES Restrooms, drinking water.

Lemon Reservoir

BOAT RAMP One concrete ramp is on the east side of the reservoir.

FISH Rainbow and brown trout and kokanee. Typical baits from shore such as Power Bait, salmon eggs and nightcrawlers will produce fair to good action for rainbow trout. Trolling flashers with assorted lures such as Mepps, Roostertails or Rapalla will produce action for trout. Trolling flashers with Cherry Bobber lures will produce kokanee.

RECREATION Fishing, boating, camping.

CAMPING There are three campgrounds: Transfer Park is five miles north of the lake (25 sites, units up to 35 feet). Near the Florida River four miles north of the reservoir is the Florida Campground (20 campsites, units up to 35 feet, and a group campground with a capacity of 100). The Miller Creek Campground is adjacent to the reservoir (12 sites, units up to 35 feet). All campgrounds include picnic tables, fire grates, vault toilets and drinking water.

CONTROLLING AGENCY San Juan National Forest.

INFORMATION Columbine Ranger District (970) 884-2512.

MAP REFERENCES San Juan National Forest map. Benchmark's *Colorado Road & Recreation Atlas:* p. 122.

Lonetree Reservoir

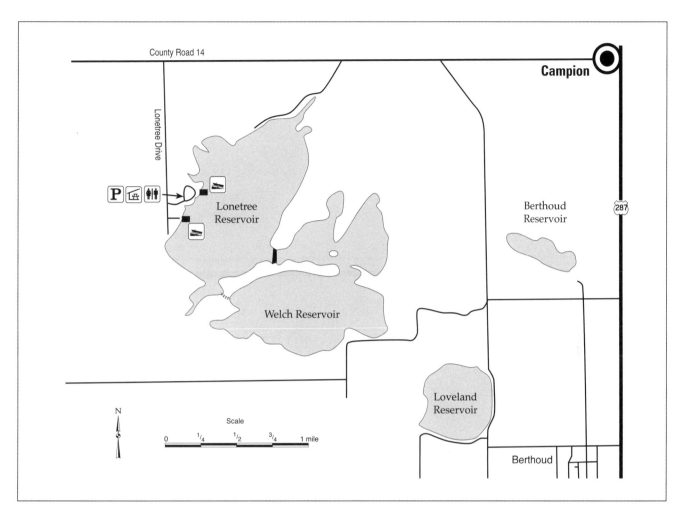

DIRECTIONS From Campion 3.5 miles west on County Road 14 to Lonetree Drive, then go one mile south to the access road to the reservoir.

FEE Habitat stamp required.

SIZE 502 acres.

ELEVATION 5,131 feet.

FACILITIES Restrooms, shade shelters.

BOAT RAMP Two ramps. No whitewater wake or sailing.

FISH Walleye, crappie, wiper, largemouth bass, channel catfish and perch. All bass in possession must be 15 inches or longer. Consult Colorado DOW fishing regulations. Fishing prohibited from boats November 1 through last day of waterfowl season. Private land surrounds most of the lake, so fishing from shore can be difficult.

RECREATION Fishing, hunting.

CAMPING Camping is prohibited.

CONTROLLING AGENCY Colorado Division of Wildlife.

INFORMATION Ft. Collins Office (970) 472-4300.

MAP REFERENCES Benchmark's *Colorado Road & Recreation Atlas:* p. 63.

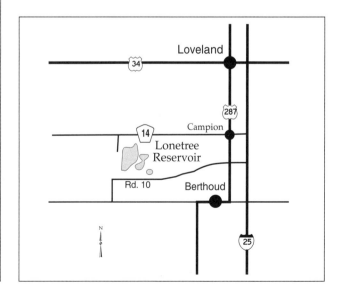

Mancos State Park

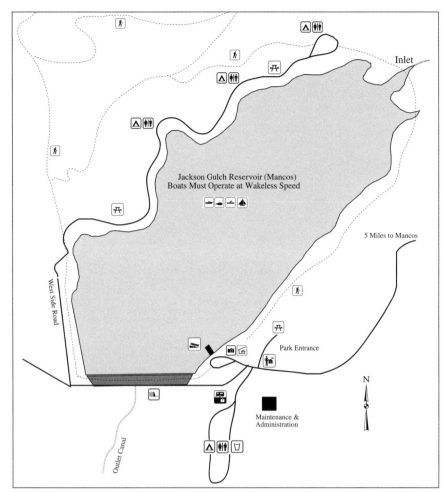

Jackson Gulch Reservoir (Mancos)
Boats Must Operate at Wakeless Speed

West Side Road

Outlet Canal

Inlet

5 Miles to Mancos

Park Entrance

Maintenance & Administration

N

do not offer electrical hookups. Along the northwest side of the reservoir there are nine campsites primarily for tent campers withrestrooms but no drinking water. Two yurts are available for rent year-round.

FISH Stocked rainbow trout. Yellow perch are also plentiful.

RESTRICTIONS Wakeless boating only, no waterskiing or swimming allowed.

OTHER INFORMATION Scenic Mancos State Park is located only 12 miles from historic Mesa Verde National Monument in southwest Colorado. Situated at an elevation of 7,800 feet on the San Juan Skyway Scenic Byway, Mancos State Park is surrounded by the majestic San Juan mountain range. The area has more than 300 land acres and the reservoir, which is often referred to as Jackson Gulch, provides 216 surface acres of water for recreation.

Jackson Gulch Dam in Mancos State Park was constructed in 1948 by the Bureau of Reclamation. It supplies the drinking and irrigation water for Mesa Verde and the surrounding rural Mancos Valley area.

There is a beautiful campground here, nestled within a nature ponderosa pine forest. Wakeless boating and excellent year-round fishing await the recreationist looking for a relaxing day. Picnickers and hikers will find sites and trails sure to please.

In addition to Mesa Verde, visitors to the area have the opportunity to view prehistoric Anasazi Indian ruins at the Anasazi Heritage Center, located in Dolores, approximately 20 miles northwest of Mancos. The Durango-Silverton Narrow Gauge Railroad in nearby Durango offers the visi-

DIRECTIONS From Mancos drive north on Colorado 184 approximately one mile to County Road 42. Drive north five miles to park.

FEE Daily or annual park pass.

SIZE 216 surface acres.

ELEVATION 7,285 feet.

FACILITIES RV dump station, drinking water, ampitheatre, park office, trails, picnic areas.

BOAT RAMP One ramp located on the south shore near the dam.

RECREATION A five mile trail system weaves through the park. The Chicken Creek Trail connects with a network of trails in the national forest. The trials are for hiking, horseback riding and mountain bikes. The park has a trailhead connecting to the Colorado Trail. Winter sports are cross county skiing, ice fishing and snowmobiling.

CAMPING The park has two campgrounds with 32 campsites in total. Main and West Campgrounds offer vault toilets, fire pits, picnic tables, and drinking water. The sites

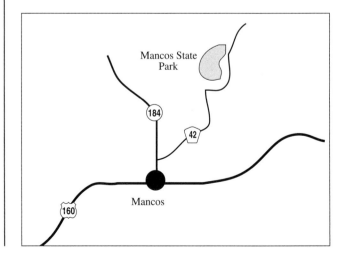

Mancos State Park

184

42

Mancos

160

Mancos State Park

tor a scenic trip through remote wilderness areas of the San Juan National Forest. Durango is 27 miles east of Mancos.

Mancos State Park is managed by Colorado State Parks in cooperation with the US Bureau of Reclamation and the Mancos Water Conservancy District.

The pleasant climate and many diversions make Mancos a not-to-be-missed stop on your next trip through southwestern Colorado.

CONTROLLING AGENCY Colorado State Parks.

INFORMATION Mancos Office (970) 533-7065.

MAP REFERENCES Benchmark's *Colorado Road & Recreation Atlas:* p. 121.

McPhee Reservoir

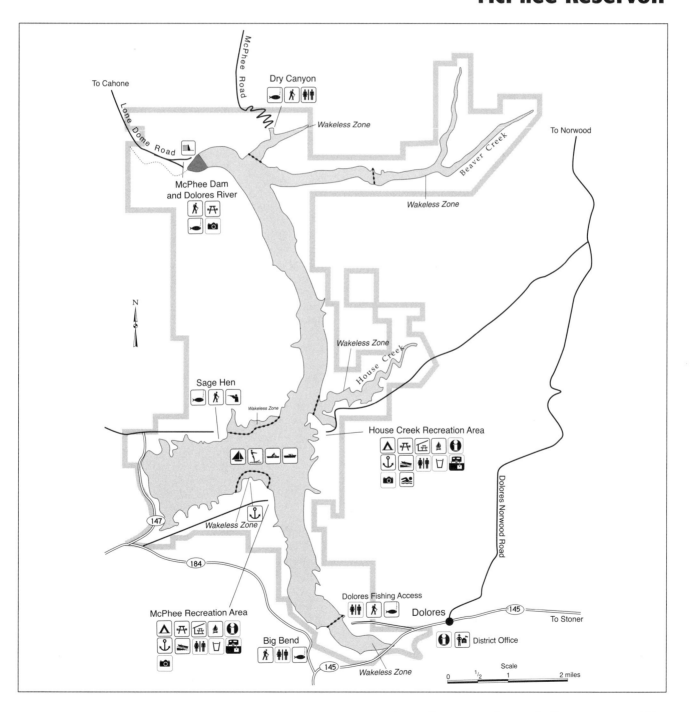

McPhee Reservoir

DIRECTIONS From Denver, take I-25 south to Walsenburg. From Walsenburg west on US 160 to Mancos. From Mancos on Colorado 184 to Dolores. Colorado 184 takes you to the reservoir.

FEE Camping fee.

SIZE Surface area is 4,470 acres.

ELEVATION 6,924 feet.

MAXIMUM DEPTH 270 feet, at the dam.

FACILITIES Drinking water, restrooms, picnic area, trailer dump station, and fish cleaning station. Near town of Dolores.

BOAT RAMP The McPhee Recreation Complex, 8 miles north of Dolores on Colorado 184, has a six-lane, concrete boat ramp. The House Creek Recreation Complex, 15 miles north of Dolores on Forest Road 526, has a four-lane boat ramp. There are fishermen access points with parking at Dolores, Dolores Town Park, Big Bend, Old McPhee Road, Sage Hen, and Dry Canyon. Car top boats can be launched from these spots.

FISH McPhee Reservoir, one of Colorado's dual fisheries, has a population of McConaughy rainbow trout and kokanee averaging 12-16 inches and a good population of smallmouth and largemouth bass, bluegill, crappie, perch, catfish. The reservoir also supports such native species as green sunfish, bullhead, and yellow perch. The best time to catch trout and kokanee is April and early May. To catch kokanee, use a Kokanee Killer and assorted lures while trolling, Power Bait or nightcrawlers while bait fishing. The best area for bass is the Beaver Creek inlet and the Sage Hen fisherman access area, around shallow sunken debris. Bass under 15 inches in length must be released immediately.

RECREATION Boating, fishing, camping, water sports.

CAMPING McPhee Recreation Complex has a 76-unit campground. The House Creek Recreation Complex has a 60-unit campground. Both sites will accommodate units up to 50 feet. McPhee Recreation Complex has drinking water, flush toilets, picnic area, trailer dump station and fish cleaning station. The House Creek Recreation Complex has composting toilets, drinking water, a picnic area and a dump station. Both areas have a limited number of electric hookups.

SPECIAL RESTRICTIONS High speed boating is restricted to the main body of the reservoir. Swimming is not allowed in the boat launching and docking areas. Waterskiing is restricted to the main body of the reservoir. Reservoir has six wakeless zone areas marked with buoys; Dry Canyon, Beaver, Sage Hen, House Creek, Dolores and the marina area. These areas are patrolled for enforcement of the wakeless areas.

Hunting and the discharge of firearms is not permitted within developed recreation sites. Hunter camps outside of developed sites are not permitted within the management area. Motorized vehicle use is not permitted with wildlife management areas from December 1 through April 30. This restriction is necessary to minimize harassment of wintering deer and elk in these areas.

OTHER INFORMATION McPhee Reservoir is located in the heart of some of the most scenic and recreation-oriented lands in the southwestern United States. A short distance to the south of McPhee Reservoir lies internationally known Mesa Verde National Park with its prehistoric Indian ruins dating to AD 500-1300.

The earliest known use of the area surrounding McPhee Reservoir began with the ancient Anasazi ("Old Ones") Indians who inhabited the mesas above the Dolores River between AD 500 and 1300. Use continued through the historic period and in 1776, Franciscan priest-explorers, Father Dominquez and Escalante traveled through the area and recorded the Dolores River country.

The town of Big Bend, established in 1880, was the forerunner to the present town of Dolores, which was established in 1891. The Big Bend townsite is inundated by the reservoir as well as the townsite of McPhee. McPhee, located about five miles downstream from Dolores, developed around a sawmill constructed in 1924. The mill finally closed down in 1948 with little remaining at present but foundations.

The Four Corners area is rich in Indian ruins and sites of the Anasazi. The Bureau of Reclamation has inventoried over 600 sites in the vicinity of the Dolores Project area. The sites vary in extent, with surface indications including rock chips, pottery, pit houses, and remains of pueblos and other structures. The sites represent past occupations of the area from perhaps 2500 BC until historic times. The entire McPhee Reservoir Project area has been recommended for nomination to the National Register of Historic Places.

CONTROLLING AGENCY San Juan National Forest.

INFORMATION Dolores Ranger District (970) 882-7296.

HANDICAPPED ACCESS The marina is fully handicapped accessible. Parking, restrooms, fishing pier and restaurant. Campgrounds have a reserved site in each loop near accessible restrooms.

MAP REFERENCES San Juan Forest map. Benchmark's *Colorado Road & Recreation Atlas:* p. 120.

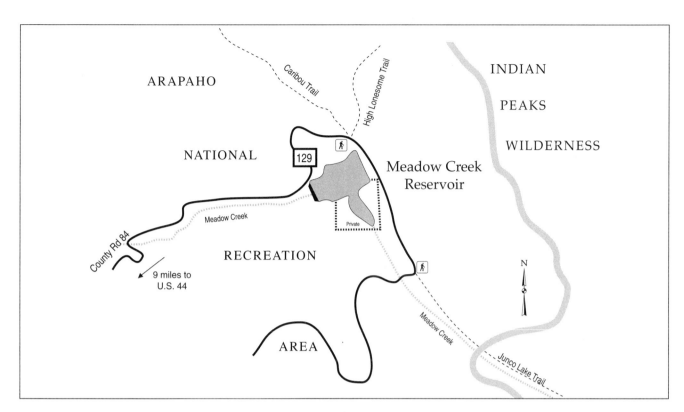

DIRECTIONS From Fraser travel north one mile. Before Tabernash, turn right onto County Road 84 along Meadow Creek. Follow County Road 84 nine miles to the reservoir.

FEE No.

SIZE 50 acres.

ELEVATION 9,990 feet.

FACILITIES Vault toilet.

BOAT RAMP No ramps. The reservoir is restricted to non-motorized boats, hand-powered craft only.

RECREATION Fishing, camping.

FISH Rainbow, brook, cutthroat, and brown trout. Typical baits such as Fireballs, cheese, marshmallows, Power Bait and nightcrawlers work well from shore. The best shore fishing usually is from the north or east shorelines and near the dam. Typical lures: Mepps, Blue Fox or Thomas lures. The best action on lures is in the early morning or late afternoon. The water level tends to fluctuate with the draw down on the reservoir. Don't let lower water levels fool you; some of the best fishing tends to be in the fall as water levels drop. You also may fish on Meadow Creek, below the dam, and chances are you'll find good fishing for brook trout for about a mile. The south shore is located on private land.

CAMPING Open camping is available near the reservoir but on undeveloped sites. Remember to clean the area and leave the site in good condition for the next person.

OTHER INFORMATION Meadow Creek Reservoir, located in the Arapaho/Roosevelt National Forest just south of Rocky Mountain National Park, offers a beautiful mountain setting and is a short drive from the towns of Fraser and Tabernash.

CONTROLLING AGENCY Arapaho & Roosevelt National Forests.

INFORMATION Sulphur Ranger District (970) 887-4100.

MAP REFERENCES Benchmark's *Colorado Road & Recreation Atlas:* p. 62.

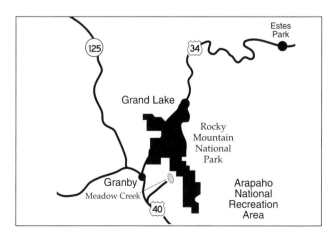

Meredith Reservoir

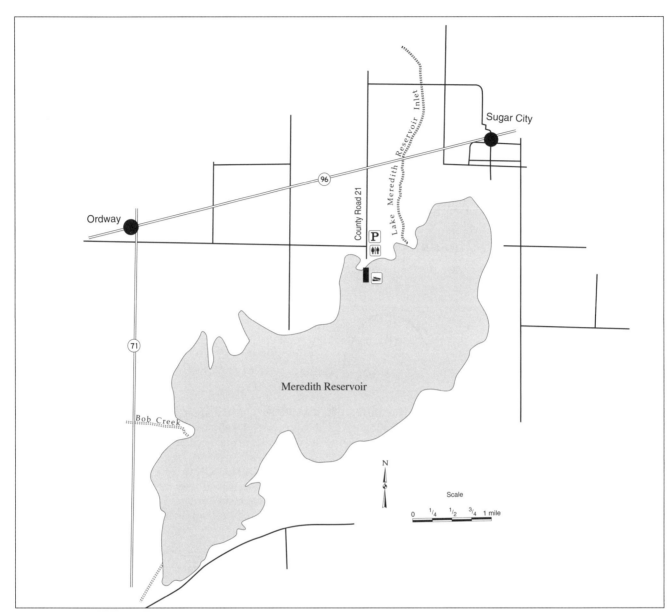

Ordway

Sugar City

96

County Road 21

Lake Meredith Reservoir Inlet

P

Meredith Reservoir

71

Bob Creek

N

Scale

0   1/4   1/2   3/4   1 mile

Meredith Reservoir

DIRECTIONS From Ordway travel three miles east on Colorado 96, then head south on County Road 21 to the reservoir.

FEE No.

SIZE 3,220 acres.

ELEVATION 4,254 feet.

MAXIMUM DEPTH 20 feet.

FACILITIES Restrooms, boat ramp.

BOAT RAMP Ramp on north side of reservoir.

FISH Walleye, bluegill, tiger muskie, wipers, channel catfish. Water drawdowns during dry years can affect the fishing.

RECREATION Fishing, waterskiing, personal watercraft, sailing, hunting, camping, wildlife observation, photography.

CAMPING Dispersed camping.

RESTRICTIONS
1. Public access is prohibited as posted from Nov. 1 through the last day of migratory waterfowl season, except to retrieve downed waterfowl.

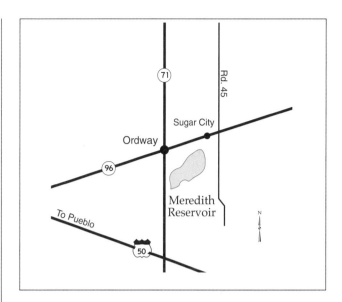

2. Wildlife Boundary is 50 feet above high water mark.
3. Access is at boat ramp area only. Land surrounding reservoir is otherwise private property.

CONTROLLING AGENCY Colorado Division of Wildlife.

INFORMATION Colorado Springs Office (719) 227-5200.

MAP REFERENCES Benchmark's *Colorado Road & Recreation Atlas:* p. 103.

Montgomery Reservoir

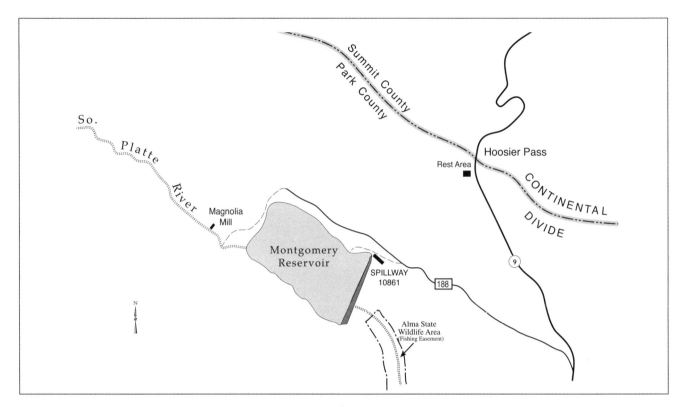

DIRECTIONS From Denver, travel south on US 285 to Fairplay. From Fairplay, travel northwest on Colorado 9 toward Hoosier Pass. About one mile before you get to Hoosier Pass, travel west on Forest Road 408 to the reservoir.

FEE No.

SIZE 97 surface acres.

ELEVATION 10,840 feet.

FACILITIES Restrooms.

BOAT RAMP No boating is allowed.

FISH Rainbow trout, catchable-size stocked, and occasional brook or brown trout. Fishing is prohibited December 1 through May 31. Fishing also is prohibited on the south side of the reservoir and from the west face of the dam, as posted. The best action generally is after June 1. Better fishing generally is found in the spring and fall. Note that algae bloom in the summer hampers angling. The best action from shore is usually with salmon eggs, Power Bait and worms rigged to suspend just off the bottom. Fair to good action on Kastmaster, Mepps and Rooster Tail lures cast from shore in the early mornings and late evenings.

OTHER INFORMATION On a four-wheel drive road, 3.5 miles west of Montgomery Reservoir, are the Wheeler Lakes (Elevation: 12,500 feet and 12,180 feet; Size: four and 28 acres). Both offer fair fishing for small cutthroat trout. Kite Lake at the Forest Service Campground also can offer fair fishing for small cutthroat trout.

RECREATION Fishing.

CAMPING No camping is allowed at the reservoir. The closest available camping is at the Alma State Wildlife Area. The Alma State Wildlife area is 1.5 miles north of Alma and west of Colorado 9 on County Road 4. The only facility is a vault toilet.

CONTROLLING AGENCY Pike & San Isabel National Forests.

INFORMATION: South Park Ranger District (719) 836-2031.

MAP REFERENCES Pike National Forest Map. Benchmark's *Colorado Road & Recreation Atlas:* p. 73.

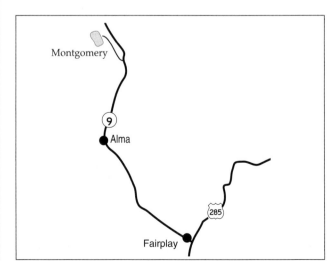

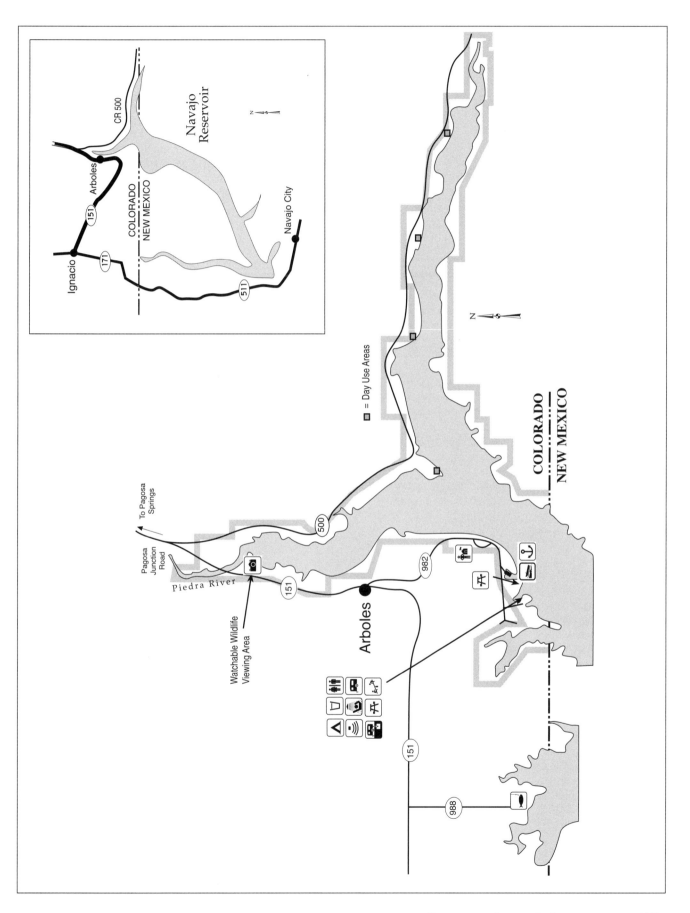

CR 500

Arboles

151

171

Ignacio

COLORADO
NEW MEXICO

Navajo
Reservoir

Navajo City

511

To Pagosa
Springs

Pagosa
Junction
Road

Piedra River

Watchable Wildlife
Viewing Area

500

151

982

Arboles

= Day Use Areas

COLORADO
NEW MEXICO

151

988

Navajo State Park

DIRECTIONS   From Pagosa Springs travel west on US 160 to Colorado 151, then go south on Colorado 151 to the reservoir.

FEE   Daily or annual pass.

SIZE   15,610 surface acres when full.

ELEVATION   6,086 feet.

MAXIMUM DEPTH   110 feet (400 feet at dam).

FACILITIES   Marina, dump station, trails, visitor center.

BOAT RAMP   Two in New Mexico. A third in Colorado, San Juan Marina at Arboles.

FISH   Warm and cold water species, with largemouth and smallmouth bass, northern pike, crappie, bluegill, channel catfish, rainbow, brown trout and kokanee salmon. Maintained almost entirely by New Mexico. Two major arms, Los Pinos River and San Juan River. Several productive side canyons in New Mexico. Access roads off Colorado 151 and County Road 500 to Pagosa Junction give access to good fishing spots on the Piedra and San Juan Rivers. Visitors planning to fish in New Mexico's waters as well as Colorado's must have fishing licenses from both states.

RECREATION   Boating, hiking, biking, swimming, sailing, personal watercraft, and other water sports.

CAMPING   There are 118 campsites, showers and flush toilets. Tiffany Campground has 25 RV sites, most of which will accommodate RVs up to 40 feet. Carracas Campground has 41 RV sites with showers and electrical hookups. Rose Campground has 39 RV sites with full hookups, plus 8 tent sites. There are 19 primitive sites are open year-round, but cannot be reserved. Many sites have pull-throughs, all can accommodate tents, trailers, or pickup campers.

OTHER INFORMATION
Navajo Reservoir, on the southwest border of Colorado, is the main attraction of Navajo State Park. The reservoir is 35 miles long, extends well into New Mexico and is in an area that is unpolluted and sparsely populated.

Navajo Dam, located in New Mexico, was constructed on the San Juan River by the US Bureau of Reclamation in 1962. The part of the reservoir situated on the Colorado side of the state line has been administered by the Colorado Division of Parks and Outdoor Recreation since 1964.

The park's 3,000 acres offer a challenge to the angler and unlimited pleasure to the boater and water-skier. Navajo boasts Colorado's largest boat ramp—80 feet wide and a quarter mile long. A marina is open April through October to provide gas, boat repairs, food service and groceries. Slip and buoy rental is available, as well as fishing and skiing gear, boat and slip rentals.

Boaters are subject either to Colorado boating statutes and regulations or to New Mexico laws, depending on which side of the state line they are on. The two states have a reciprocal agreement honoring current boat registrations, and boat registrations are available at the visitor center.

Waterfowl, shorebirds, birds of prey, including bald eagles, and songbirds such as waxwings, thrushes and meadow larks, inhabit the area along with game birds like dove, grouse and wild turkey. Beaver, mink, fox, deer, elk, and rabbits may be seen, and in the remote areas, coyotes, bobcat and mountain lion. Hunters come here during established seasons for deer, elk and bird hunting.

CONTROLLING AGENCY   Colorado State Parks.

INFORMATION   Arboles Office (970) 883-2208. Fishing condition reports are available at the visitor center.

MAP REFERENCES   Benchmark's *Colorado Road & Recreation Atlas:* p. 123.

North Sterling State Park

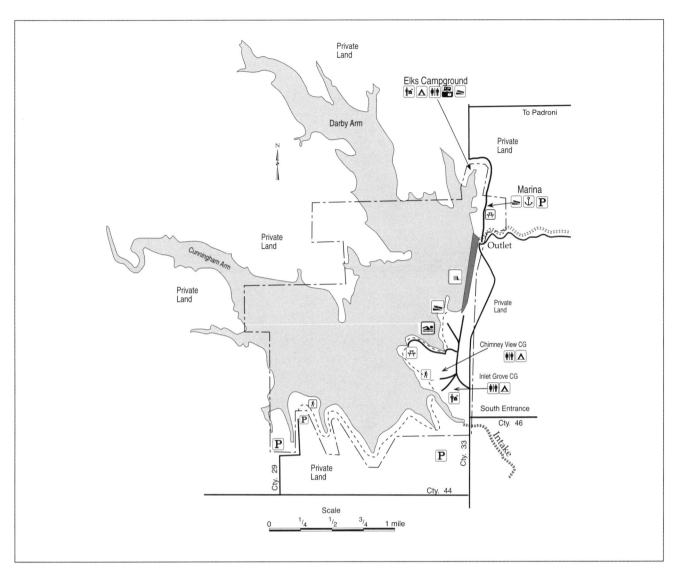

DIRECTIONS From Sterling at Colorado 14, travel 8 miles north on 7th Avenue to County Road 46. Then proceed two miles west to County Road 33, and go one mile north to the reservoir.

FEE Daily or annual pass.

SIZE 2,880 acres.

ELEVATION 4,050 feet.

FACILITIES Restrooms, ampitheatre, park office, playground, trails, visitor center, dump station, marina, picnic pavilion, boat ramps.

BOAT RAMP There are three ramps located along the east side of reservoir near County Road 33.

FISH Walleye, crappie, perch, bass, tiger muskie, bluegill, catfish and wiper. The reservoir is regarded as one of the top warm-water fisheries in the state, particularly for walleye and wiper. The reservoir is open to fishing year-round. The area is closed to boating November 1 through the end of waterfowl hunting season. The water level fluctuates during the irrigation season. Boaters are warned to be alert for submerged hazards.

RECREATION Fishing, swimming, biking, hiking, horseback riding, waterskiing, personal watercraft, sailing, hunting.

CAMPING There are 50 campsites north of the marina at Elks Campground that are open year-round. All sites at Elks Campground have a table, fire ring, shade shelter, and electricity. A centrally located camper services building offers showers, flush toilets, and laundry. A dump station is located at the entrance to Elks Campground. There are 91 sites are available at the Inlet Grove and Chimney Views Campgrounds. The sites share a camper services building. Inlet Grove has electrical hookups, while Chimney View does not.

North Sterling State Park

CONTROLLING AGENCY Colorado State Parks.

INFORMATION Sterling Office (970) 522-3657; Marina (970) 522-1511.

RESTRICTIONS Private property is adjacent to much of the reservoir. Please camp in designated areas only.

North Sterling Park is open daily year-round. During the winter months, the park offers camping, hunting, ice fishing and opportunities for photography and wildlife observation. Snowmobiling is not permitted.

OTHER INFORMATION Majestic bluffs and expansive views of the nearby high plains welcome visitors to North Sterling Reservoir.

North Sterling Reservoir is a boater's paradise, offering a 3,000-acre lake with an interesting array of coves and fingers to explore. The surrounding area is home to a diversity of wildlife and has a rich historical heritage.

Built during the turn of the century, the reservoir serves as an important irrigation facility along the lower South Platte River Valley.

Colorado State Parks acquired the area in 1992 and manages the reservoir through a perpetual easement with North Sterling Irrigation District.

HANDICAPPED ACCESS Handicapped-accessible visitors center, showers, restrooms, hunting, picnic tables, camping and fishing. Some facilities accessible with assistance.

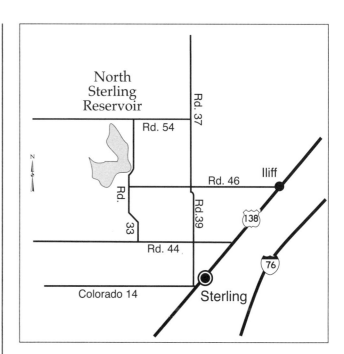

MAP REFERENCES Benchmark's *Colorado Road & Recreation Atlas:* p. 52.

Paonia State Park

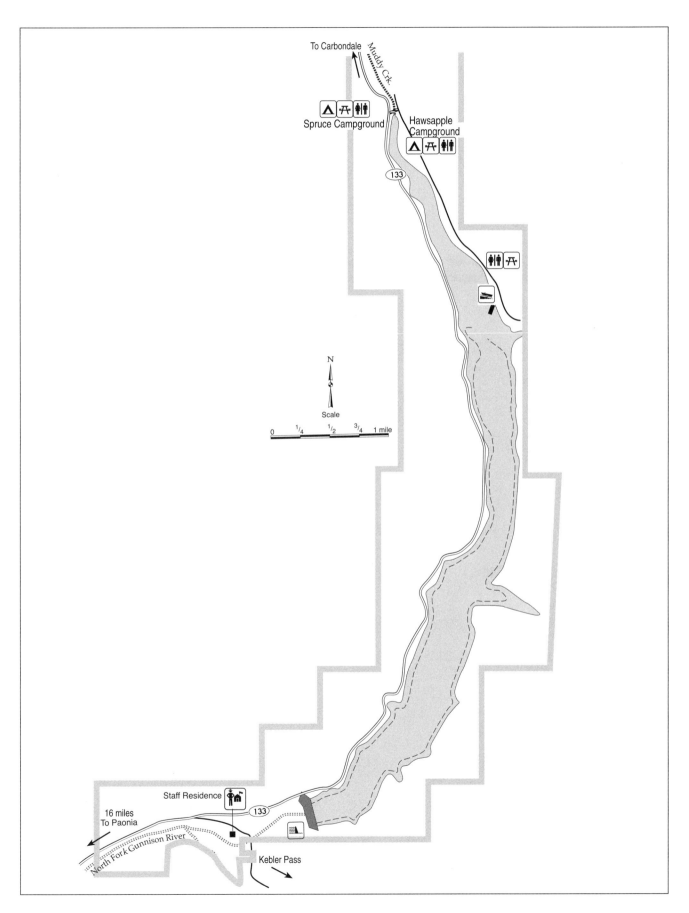

To Carbondale

Muddy Crk.

Spruce Campground

Hawsapple Campground

133

N

Scale

0   ¼   ½   ¾   1 mile

Staff Residence

16 miles
To Paonia

133

North Fork Gunnison River

Kebler Pass

Paonia State Park

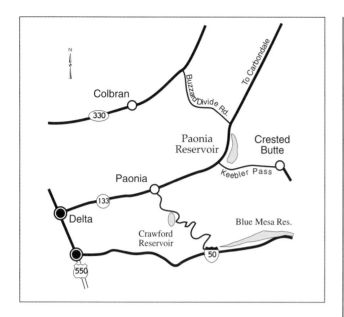

DIRECTIONS The reservoir is 16 miles east of Paonia on Colorado 133.

FEE Daily or annual pass. Camping fees.

SIZE 334 acres.

ELEVATION 6,450 feet.

MAXIMUM DEPTH 120 feet.

FACILITIES Restrooms, picnic tables.

BOAT RAMP Ramp is located on the east side of reservoir.

FISH Rainbow trout, northern pike, suckers. Mid-summer best for pike fishing.

RECREATION Fishing, waterskiing, personal watercraft, sailing.

CAMPING There are 13 campsites in two campgrounds— Spruce and Hawsapple. Both have tables, grills and vault toilets. No drinking water is available.

CONTROLLING AGENCY Colorado State Parks.

INFORMATION Crawford Office (970) 921-5721.

MAP REFERENCES Benchmark's *Colorado Road & Recreation Atlas:* p. 85.

Platoro Reservoir

DIRECTIONS From Mount Vista, travel 12 miles south on Colorado 15 to Forest Road 250. Take Forest Road 250 for about 30 miles to Forest Road 247. Turn west on Forest Road 247 to Platoro Reservoir, about 1.5 miles.

FEE Camping fees.

SIZE 700 acres.

ELEVATION 9,970 feet.

FACILITIES Restrooms.

BOAT RAMP Primitive amp on north side of reservoir.

FISH Brown and rainbow trout. Rated good fishing despite water level variations.

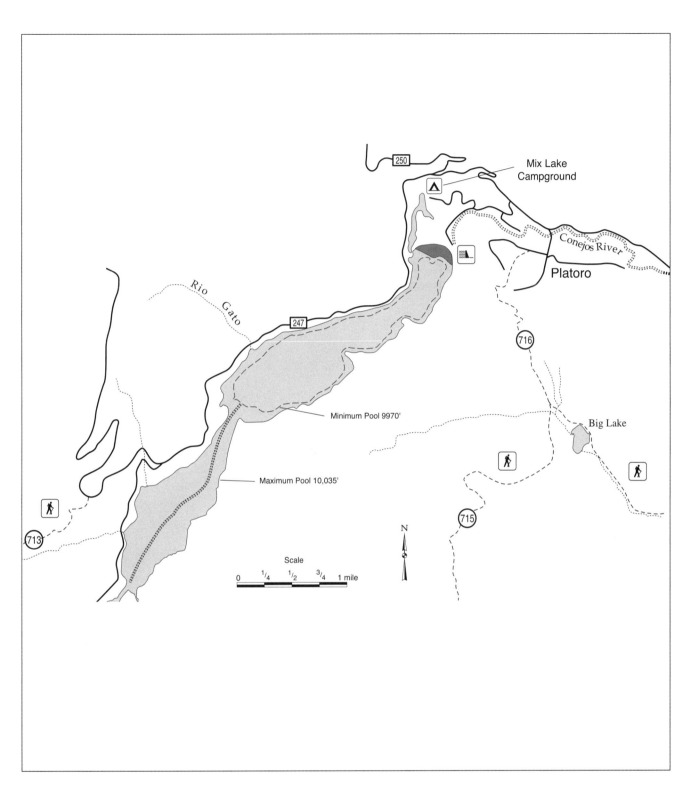

RECREATION Boating, fishing.

CAMPING Mix Lake Campground has 22 sites.

CONTROLLING AGENCY Rio Grande National Forest.

INFORMATION Divison of Wildlife, Monte Vista Office (719) 587-6900.

MAP REFERENCES Rio Grande National Forest. Benchmark's *Colorado Road & Recreation Atlas:* p. 124.

Queens State Wildlife Area

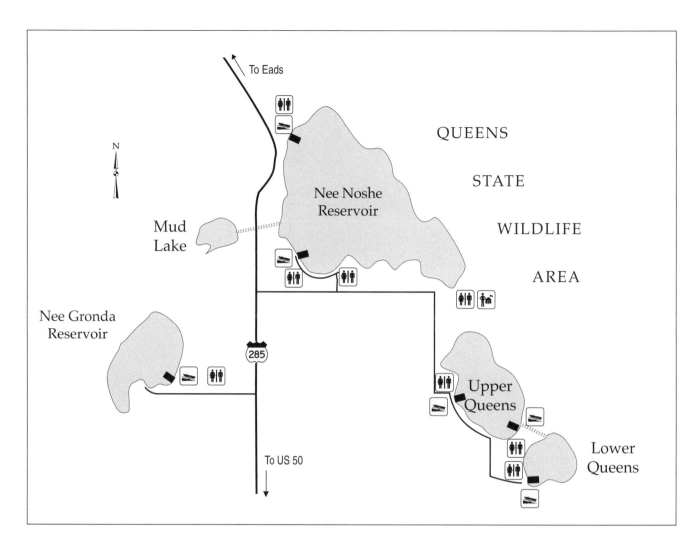

Queens State Wildlife Area

DIRECTIONS From Denver, go east on I-70 to Limon, south from Limon on Colorado 287 to Eads, then south from Eads 15 miles on Colorado 287 to County Road C. Go 1 mile west to the State Wildlife Area.

FEE No.

SIZE Varies greatly. Nee Gronda—3,490 acres, Nee Noshe—3,696 acres when full.

ELEVATION 3,876 feet.

FACILITIES Restrooms, boat ramps.

BOAT RAMP At Nee Gronda there is a gravel ramp on the east side that is good for larger boats. It requires backing in to the lake to launch boats. Nee Noshe has boat ramps suitable for small and medium boats. Launching is difficult with low water level. Call park office for status.

FISH Warm water fishing at Queens SWA is rated good. Nee Gronda; Fishing for wiper, crappie, catfish, drum, walleye, saugeye, and white bass. Wiper are caught with minnows, nightcrawlers, carp bait. Hoten Tots, Thin Fins and Rattletrip lures near the bottom. Crappie are caught on white jigs and minnows around the sunken trees and brush. Catfish are caught on carp bait, shrimp and nightcrawlers on the bottom during the day, best fishing in the evening. Walleye and saugeye are caught on Wall Bangers with a nightcrawler worm, harness, minnows and rubber worms in color of motor oil. Watch for evening thunderstorms and high winds. No wake in Lower Queens and in the channel between upper and lower Queens. Use caution when boating, if storms are building, get off the lake as soon as possible. Note: In dry years the lakes can be very low or empty. Call in advance about reservoir water levels.

RECREATION Boating, fishing, swimming, hunting, sail boarding, waterskiing, hunting, wildlife observation and photography.

CAMPING On the east side of the Nee Gronda reservoir, there is open camping with no fee. Pit toilets available but there is not any other facilities. In the Cottonwood Park area improved camping sites is available for a fee. Showers and restrooms are available. Camping is in typical eastern

Queens State Wildlife Area

plains terrain, wide open and flat with very few trees for shade. No potable water is available.

SPECIAL RESTRICTIONS

1. Waterfowl hunting is prohibited except when each hunter is properly registered in the check-in-point.
2. Fishing and boating are prohibited from November 1 through the last day of the regular migratory waterfowl season.
3. Hunting and trapping are controlled, as posted, during the regular migratory waterfowl season including any periods between split seasons. All hunters must check in and out at the check station daily during the goose hunting season. Pheasant hunting is restricted to no more than 30 hunters at a time during the goose hunting season. Pheasant hunting is from noon to sunset.
4. Boating in such a manner as to create a white water wake is prohibited in the channel between North and South Queens Reservoirs.
5. Swimming is permitted only in a designated area, as posted.

CONTROLLING AGENCY Colorado Division of Wildlife.

INFORMATION Lamar Office (719) 336-6600.

MAP REFERENCES Benchmark's *Colorado Road & Recreation Atlas:* p. 105.

Quincy Reservoir

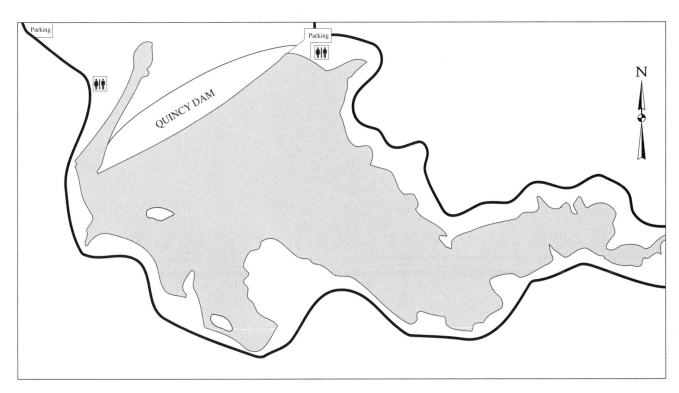

DIRECTIONS From Denver, take I-25 to I-225 to Parker Road. Then take Parker Road to Quincy Avenue, follow Quincy Avenue east to one mile past Buckley Road. The reservoir is on south side.

FEE City of Aurora daily or annual pass required for each visitor.

SIZE 161 acres.

ELEVATION 5,699 feet.

MAXIMUM DEPTH 58 feet.

FACILITIES Restrooms, handicapped fishing dock.

BOAT RAMP Dirt ramp for hand-launched boats. No gas motors are permitted. Boat rental available.

FISH Fishing for trout, large and smallmouth bass, tiger muskie and perch. Check park regulations for various bag and possession limits. Fishing by artificial flies or lures only. Scented flies and lures—either manufactured or spayed on—are not permitted. Possession or use of bait is strictly prohibited. Trout are caught with Jigs, Mepps spinners, Kastmasters or various fly patterns. Best results for trout is in the early morning or at dusk. Bass are caught in various structure areas using spinner baits, topwater baits, soft plastics and jigs. Tiger muskies are caught in a wide variety of areas with spinner baits, topwater baits, large plugs or Rapala lures. Perch are caught with Jigs in various depths of water.

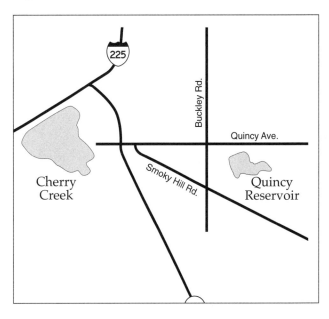

Quincy Reservoir

RECREATION Fishing, hiking, wildlife watching, boating, and jogging.

CAMPING No camping; day-use area only.

CONTROLLING AGENCY City of Aurora Parks and Open Space Department.

INFORMATION Park Office (303) 693-5463. The park is open March 1–October 31. Hours: a half-hour before sunrise to one hour after sunset.

MAP REFERENCES Benchmark's *Colorado Road & Recreation Atlas:* p. 75.

Rampart Reservoir

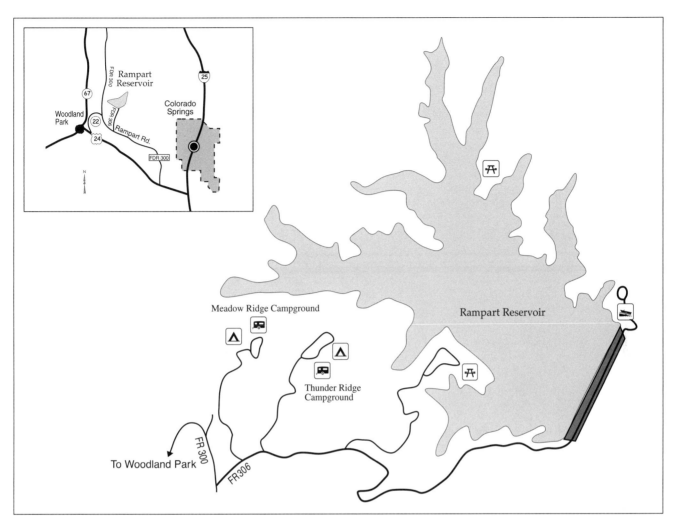

DIRECTIONS From Woodland Park travel on County Road 22 (Rampart Range Road) about three to miles Forest Road 300 (also known as Rampart Range Road). Then go south for three miles to Forest Road 306, then head east 2.5 miles to the reservoir.

FEE Fee for day-use. If camping, day-use fee included in charge.

SIZE 500 acres.

MAXIMUM DEPTH 200 feet.

ELEVATION 9,000 feet.

FACILITIES Restrooms, and picnic areas are available—one is accessible only by boat or hike and features 12 picnic sites. The second has drive-up access and features 34 sites.

BOAT RAMP Concrete ramp. No wake boating.

RECREATION Fishing. Biking and hiking trail encircles the reservoir. No swimming or other water contact sports.

CAMPING Two Forest Service campgrounds are available: Meadow Ridge offers 19 campsites, and Thunder Ridge has 21sites. The area is open for tents, truck campers and trailers 30 feet or shorter. Both campgrounds offer picnic tables, drinking water, fire pits and vault toilets. No overnight camping in the picnic areas.

FISH Rainbow and lake trout. Fishing is good for 10- to 14-inch rainbow trout using Power Bait, nightcrawlers or salmon eggs from shore. Fair to good action trolling Pop Gear with a nightcrawler, and lake trout have been caught trolling on assorted lures. Note: Bag, possession and size limits on lake trout is one fish 20 inches or longer.

CONTROLLING AGENCY Pike & San Isabel National Forests.

INFORMATION Pikes Peak Ranger District (719) 636-1602.

MAP REFERENCES Pike National Forest. Benchmark's *Colorado Road & Recreation Atlas:* p. 89.

Red Dirt Reservoir

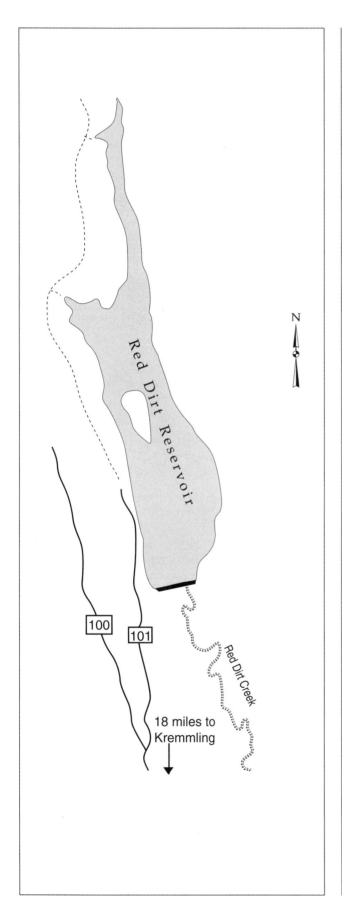

N

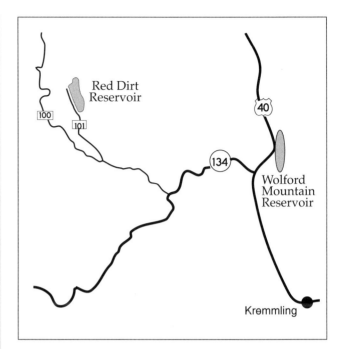

DIRECTIONS From Denver, travel west on I-70 to Dillon, then travel north from Dillon on Colorado 9 to Kremmling. Travel northwest from Kremmling on Colorado 134 then turn right on Forest Road 100 and travel about two miles to the reservoir.

FEE No.

SIZE 130 surface acres.

ELEVATION 9,055 feet.

FACILITIES Restrooms.

BOAT RAMP No boat ramps. Electric motors or canoes only.

FISH Rainbow trout; catchable size stocked. The best action generally has been by fishing from the shoreline with bait kept suspended just off the bottom. Berkley Power Bait in pink or green, inflated nightcrawlers, salmon eggs or corn rigged just off the bottom tends to produce fair to good results. Casting lures, such as Mepps or Rooster Tails, from shore will at times produce fair results. Fly fishing tends to be better in the early morning or late afternoon, when the trout are feeding on the surface. The reservoir is under private ownership but is open to fishing through an agreement with the US Forest Service. Public fishing access is allowed from the reservoir's entire shoreline.

RECREATION Fishing, hiking, mountain biking.

CAMPING Camping is allowed along the west shoreline within the US Forest Service boundary, though no camping

Red Dirt Reservoir

sites are designated. Camping must be 100 feet away from the water. Please remember to take out everything you bring in. Be aware the road past the reservoir tends to be rough.

The east shoreline is primarily open with some sagebrush and little shade. The west shoreline features some evergreen and aspen growth for shade and tends to get brushy closer to the shoreline.

CONTROLLING AGENCY Medicine Bow & Routt National Forests.

INFORMATION Yampa Ranger District (970) 638-4516.

MAP REFERENCES Routt National Forest map. Benchmark's *Colorado Road & Recreation Atlas:* p. 60.

Red Feather Lakes

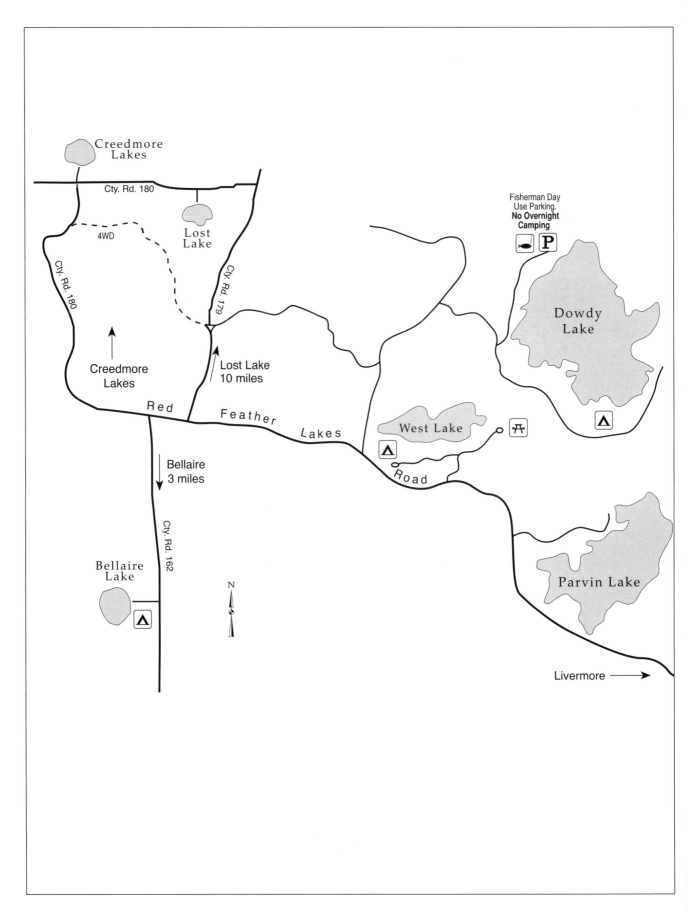

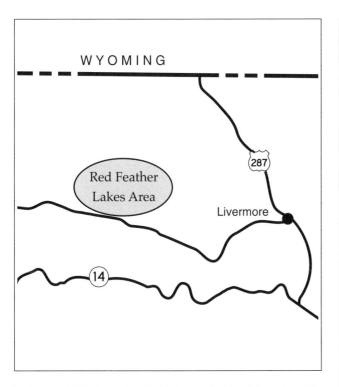

DIRECTIONS From Fort Collins on US 287, travel north 24 miles to the town of Livermore. Travel west on Red Feather Lakes Road for 27 miles to the lakes.

FEE Day-use fee. Camping fee.

SIZE Dowdy Lake, 120 acres.
West Lake, 24.9 acres.
Parvin Lake, 63 acres.
Bellaire Lake, 9.8 acres.
Creedmore Lake, 10 acres.
Lost Lake, 7.5 acres.

ELEVATION 8,365 feet.

FACILITIES Water, restrooms.

BOAT RAMPS
Dowdy Lake has a concrete boat ramp on the west side that is open for gasoline engines. It is a no whitewater-wake zone.
West Lake has a primitive gravel ramp. West Lake is open for craft propelled by hand, wind or electric motors.
Parvin Lake is open only for shore fishing or belly boats.
Bellaire Lake has no boat ramp.

SPECIAL RESTRICTIONS Contact the Colorado Division of Wildlife for boating regulations for Bellaire, Creedmore and Lost Lakes.

FISH Rainbow, brook, brown and cutthroat trout. On Dowdy and West lakes, fishing is primarily bait. Salmon eggs and Power Bait produce the best results, and you'll get good results trolling with half a nightcrawler on Pop Gear or using Mepps and Roostertail lures. On Parvin Lake, you may use only flies or lures. Possession and size limits are posted at the check station. There are special closures during the year, so check fishing regulations for information.

RECREATION Fishing, boating.

CAMPING
Bellaire Lake: Campground has 26 sites, 21 sites have electric hookups. Water, picnic tables, flush and vault toilets, and fire grills. All suitable for RVs up to a maximum length of 60 feet.
Creedmore Lake: No developed camping, tent camping only. No camping within 100 feet of water. No tables, toilets, fire grates or drinking water is available.
Lost Lake: Limited undeveloped camping. No camping within 100 feet of the lake. No drinking water, toilets, tables or fire grates are available.
Dowdy Lake: The campground offers 62 camp sites, with picnic tables, fire grates and drinking water. Vault toilets are also available.
West Lake: The campground offers 35 campsites. Water, picnic tables, vault toilets and fire grates are available.
Parvin Lake: No camping.

CONTROLLING AGENCY Arapaho & Roosevelt National Forests.

INFORMATION Canyon Lakes Ranger District (970) 295-6700.

HANDICAPPED ACCESS Handicapped-accessible facilities at Bellaire Lake; other lakes are not handicapped accessible.

MAP REFERENCES Arapaho & Roosevelt National Forests map.
Benchmark's *Colorado Road & Recreation Atlas:* p. 48.

Ridgway State Park

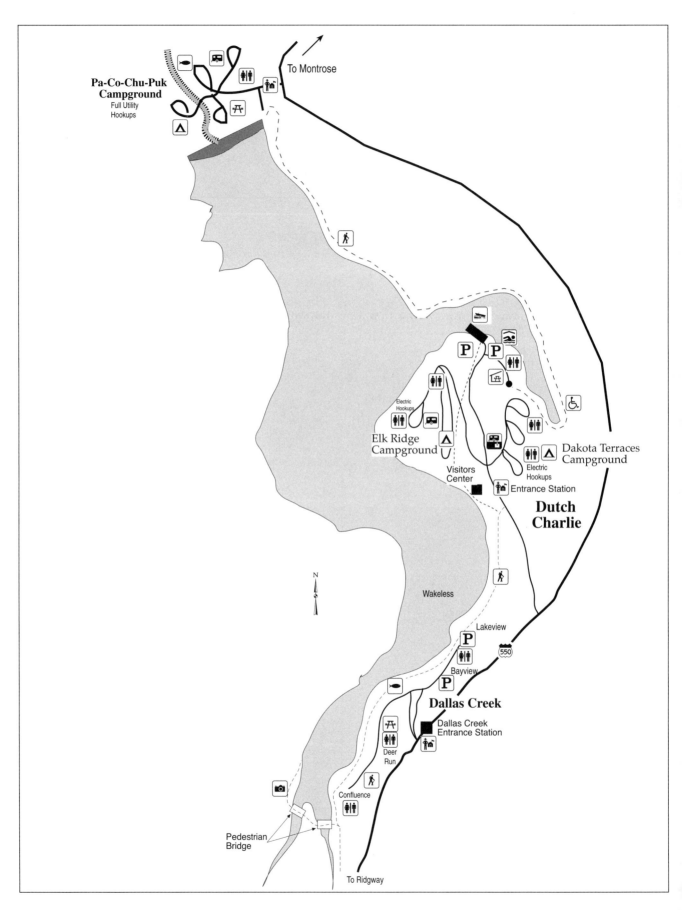

To Montrose

Pa-Co-Chu-Puk Campground
Full Utility Hookups

N

Elk Ridge Campground
Electric Hookups

Visitors Center

Entrance Station

Dakota Terraces Campground
Electric Hookups

Dutch Charlie

Wakeless

Lakeview

Bayview

550

Dallas Creek
Dallas Creek Entrance Station

Deer Run

Confluence

Pedestrian Bridge

To Ridgway

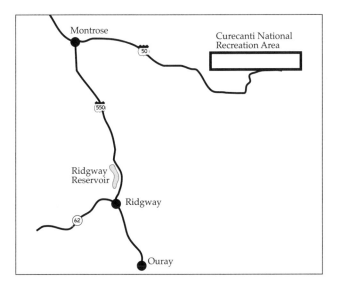

DIRECTIONS From Montrose drive 20 miles south on US 550 to the reservoir, located approximately 4 miles north of the town Ridgway. Parking is available on the west side of highway.

FEE Daily or annual pass.

SIZE 1,000 acres.

ELEVATION 6,880 feet.

MAXIMUM DEPTH 200 feet.

FACILITIES Restrooms, dump station, water, full-service marina, visitor center, swim beach, playground, fishing piers, fish cleaning station, electric hook-ups, trails.

BOAT RAMP Four-lane concrete ramp. Marina.

FISH Rainbow trout are stocked annually by the Division of Wildlife. Brown trout, which inhabited the Uncompahgre River prior to construction of the dam, are also in the reservoir. Kokanee and yellow perch can also be found.

RECREATION Fishing, boating, horseback riding, scuba diving, waterskiing, camping, swimming, sailing, surf boarding, hunting, hiking.

CAMPING Ridgway has three picturesque campgrounds. Dakota Terraces is within walking distance of the lake and swim beach, and Elk Ridge is located in a pinion-juniper forest with panoramic views of the San Juan Mountains. Pa-Co-Chu-Puk Campground on the north side of reservoir features full utility hookups. There are 283 campsites, including 25 walk-in tent areas. The campgrounds can also accommodate trailers, campers and motor homes. A camper services building with modern restrooms, hot showers and laundry facilities is also on the premises. All sites, except the tent sites, have electrical hookups, and

drinking water is available. Three yurts are available for rent in the Dakota Terraces Campground.

HIKING Fifteen miles of developed trails wind through the park connecting the various facilities. The trails range from a moderate walk to challenging hikes. The Forest Discovery Nature Trail near the visitor center is a 0.5-mile self-guided trail.

WILDLIFE AND HUNTING Winter is the best time for wildlife watching at Ridgway State Park. Deer, elk, small mammals, eagles, waterfowl, raptors and song birds can often be spotted by visitors. However, due to limited backcountry and the closeness of recreation facilities, hunting is very limited. Hunters should check with the park office for specific closures and regulations.

SWIMMING Swimming is permitted only at the Dutch Charlie swim beach. This modern facility has changing rooms with lockers and showers. Along with a refreshing swim, visitors can enjoy a rousing game of volleyball, a family picnic or a walk on the beach.

Children 12 years or younger must be accompanied by an adult. The beach does not have life guards, so swim at your own risk. There is a deluxe playground, offering hours of entertainment for children.

CONTROLLING AGENCY Colorado State Parks.

INFORMATION Ridgway Office (970) 626-5822.

MAP REFERENCES Benchmark's *Colorado Road & Recreation Atlas:* p. 96

HANDICAPPED ACCESS More than 80 percent of the facilities at Ridgway State Park are handicapped-accessible. Among the many accessible facilities the park has a wheelchair walkway leading to the swim beach and the tables and grills are constructed on elevated cement pads. For complete information on handicapped accessibility call the park office.

Rifle Gap State Park

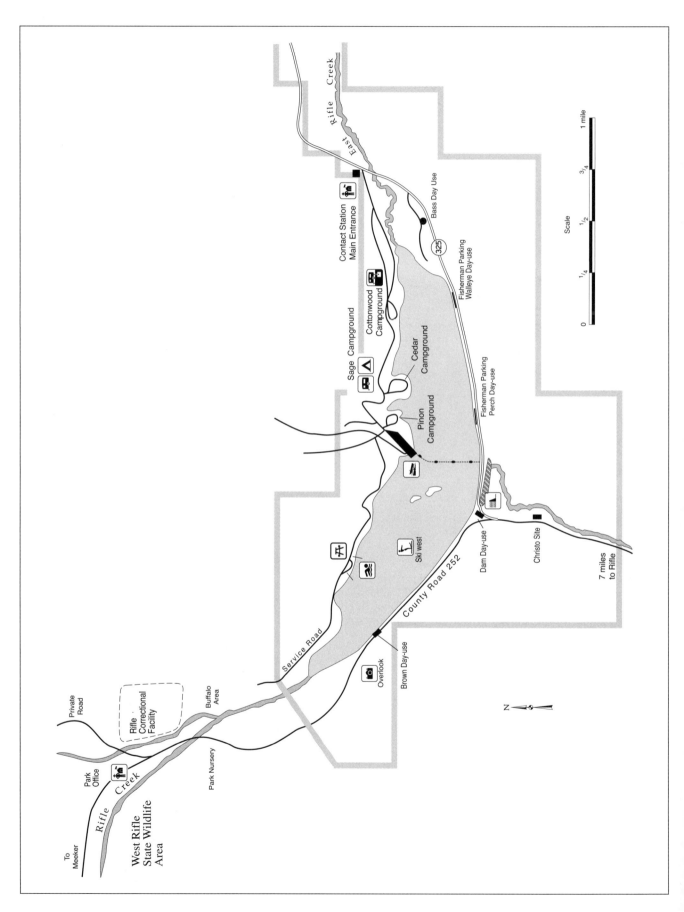

East Rifle Creek

Rifle Creek

Contact Station
Main Entrance

Bass Day Use

325

Fisherman Parking
Walleye Day-use

Sage Campground

Cottonwood
Campground

Cedar
Campground

Pinon
Campground

Fisherman Parking
Perch Day-use

Ski west

County Road 252

Service Road

Dam Day-use

Christo Site

7 miles
to Rifle

Overlook

Brown Day-use

Private
Road

Rifle
Correctional
Facility

Buffalo
Area

Park
Office

Park Nursery

Rifle Creek

To
Meeker

West Rifle
State Wildlife
Area

N

Scale

0   ¼   ½   ¾   1 mile

Rifle Gap State Park

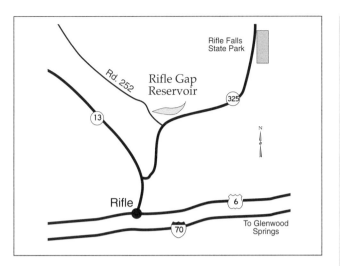

DIRECTIONS From Rifle drive seven miles north on Colorado 325 to the reservoir.

FEE Daily or annual state park pass.

SIZE 360 acres.

ELEVATION 6,000 feet.

MAXIMUM DEPTH 87 feet.

FACILITIES Full-service state park with a dump station.

BOAT RAMP Ramp located on north side of reservoir near Pinon Campground.

FISH Rainbow, brown trout, pike, crappie, bluegill, walleye, perch and smallmouth, and largemouth bass. Underwater spear fishing is permitted throughout the year. All smallmouth and largemouth bass between 12 and 15 inches must be returned to the water immediately.

RECREATION Fishing, waterskiing, personal watercraft, swimming, hunting, ice fishing, ice skating, snowmobiling.

CAMPING The park has five camping loops, providing 89 campsites that can accommodate tents, small trailers and campers. There are some pull-through sites for larger units. Cottonwood Loop has electrical hookups, restrooms, and showers. Lakeview has full hookups, picnic tables, fire pits, showers, and restrooms. Cedar has electrical hookups, vault toilets, and shared water facilities. Sage and Pinon have covered picnic tables, vault toilets, and fire pits. Camp-

Rifle Gap State Park

ground users must have a camping permit, in addition to a park pass.

OTHER INFORMATION Rifle Gap Reservoir is in northwestern Colorado near Rifle. It was constructed in 1967 by the US Bureau of Reclamation.

Although separated by a few miles, Rifle Gap Reservoir and Rifle Falls are two popular state parks administered by the Colorado Division of Parks and Outdoor Recreation. The 360-surface-acre reservoir and the falls with its mysterious caves and unique triple waterfall are unified by their beauty and superb recreational opportunities.

Rifle Gap's clean, clear waters afford some of the best scuba diving, boating, swimming and waterskiing in Colorado. Golfers can pursue their sport here, as an 18-hole golf course is located in the vacinity.

The sandstone and shale cliffs which frame the reservoir were the site of Cristo's Curtain, erected in 1972 by the internationally known artists Cristo and Jean-Claude.

Shore vegetation is pinon, juniper, grasses and sagebrush. The area around Rifle Gap and Falls is rich in history.

For campers, there are 89 sites that can accommodate tents, trailers and pickup campers. There are some pull-through sites and a dump station.

Waterskiing is popular, as well as swimming and scuba diving. In cold weather there is ice fishing. Winter also offers the fun of ice skating and snowmobiling.

Big game hunting is excellent in the White River National Forest and on lands administered by the US Bureau of Land Management, both adjacent to the park. Rifle Gap serves as a base camp area for hunters using these lands. Hunting within the park boundary is controlled.

Many small mammals ranging from beaver, chipmunks and rabbits to weasels and bobcats live around the area. Blue heron, several kinds of hummingbirds and a wide variety of ducks and other waterfowl are found here. These wild creatures might easily be seen when taking the self guided nature trail. The trail, complete with interpretive view stations, leads walkers above and over Rifle Falls.

SPECIAL RESTRICTIONS All boats must carry the following safety equipment:
1. A US Coast Guard approved life preserver for each person on board.
2. A ring buoy for boats 16 feet in length and over.
3. An effective whistle or horn; sirens are not allowed.
4. A US Coast Guard approved fire extinguisher.

CONTROLLING AGENCY Colorado State Parks.

INFORMATION Rifle Office (970) 625-1607.

HANDICAPPED ACCESS Handicapped accessible restrooms, swimming, hunting, picnic tables, camping and fishing. Some facilities accessible with assistance.

MAP REFERENCES Benchmark's *Colorado Road & Recreation Atlas:* p. 70.

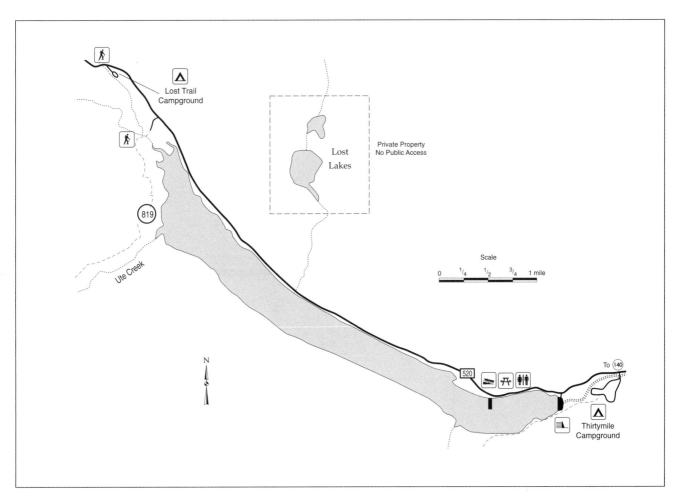

DIRECTIONS   From Creede travel 20 miles west on Colorado 149 to Forest Road 520. Turn left and continue for approximately 13 miles to reservoir.

FEE   No.

SIZE   1,500 feet.

ELEVATION   9,392 feet.

FACILITIES   Restrooms.

BOAT RAMP   Ramp located on north side of reservoir.

FISH   Brown, cutthroat, and rainbow trout. Rated good to fair. Good fishing for rainbow and cutthroat on the Rio Grande River above the reservoir.

RECREATION   Fishing, boating, picnicking. Trailhead to Weminuche Wilderness.

CAMPING   Day-use area only. However there are three campgrounds within close proximity with a total of 62 sites for tents, campers or trailers. River Hill Campground has 20 units; Thirtymile Campground has 35 units; and Lost Trail Campground has 7 units.

CONTROLLING AGENCY   Rio Grande National Forest.

INFORMATION   Divide Ranger District (719) 658-2556.

MAP REFERENCES   Rio Grande National Forest map. Benchmark's *Colorado Road & Recreation Atlas:* p. 111.

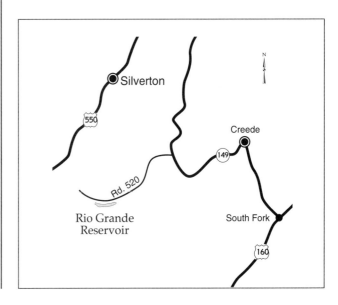

Ruedi Reservoir

Ruedi Reservoir

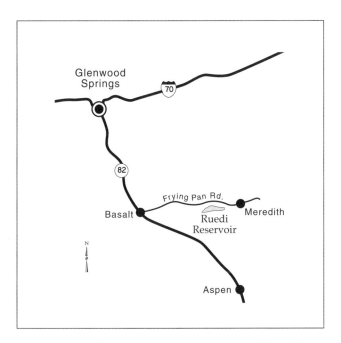

DIRECTIONS   From Denver, travel west on I-70 to Glenwood Springs. From Glenwood Springs, travel south on Colorado 82 to Basalt. Travel east from Basalt on Colorado 105 along the Frying Pan River to Ruedi Reservoir.

FEE   Camping, boat launch and parking fees.

SIZE   1,000 acres.

ELEVATION   7,800 feet.

FACILITIES   Marina, restrooms, water dump station, food store at Meredith.

BOAT RAMP   Two ramps are available. One ramp, on the inlet area, is gravel. A public boat launching ramp is located at the Ruedi Marina. It is the only ramp on the reservoir where large boats may be launched. The ramp allows boats to be launched from a water level of 7,692 feet to the high water line at 7,766 feet. A boat dock is also provided. Other amenities include drinking water, a toilet and a paved parking area.

FISH   Rainbow, lake, and brown trout. Rainbow, which average 10–14 inches, are best caught during the day using worms and salmon eggs from shore. Good results also have come from using flashers or Pop Gear trailing a nightcrawler from boats. Some large browns are caught in the late evenings trolling or shore fishing with lures and

Ruedi Reservoir

assorted fly patterns. Good fishing near the inlet in the early spring when the ice-off occurs.

RECREATION Wind sailing, boating, personal watercraft, fishing, picnicking.

CAMPING Four Forest Service campgrounds are available with a total of 81 campsites. Mollie B. Campground has 26 sites for units up to 40 feet. Little Maude has 22 sites for units to 40 feet. Little Mattie has 20 sites for units to 40 feet. Dearhamer has 13 sites for units to 35 feet. Forest Service fees vary per night depending on the site. All camp areas have running water, toilets, picnic tables and fire rings. A dump station is located between Little Maude and Mollie B. Campgrounds. Both campgrounds have flush toilets and hot water.

OTHER INFORMATION Ruedi Reservoir is located in the Fryingpan River drainage of west central Colorado. The reservoir and surrounding area are located within Pitkin and Eagle Counties approximately 13 miles east of Basalt.

The land in the Fryingpan Valley is almost exclusively rural in character, with approximately 60,000 acres designated as wilderness. The climate conditions are characterized by mild summers and cold winters. Most of the land in the reservoir area ranges in elevation from 7,400 feet to about 11,500 feet. This is the montane life zone. Stands of spruce, fir and lodgepole pine cover the steep north-facing slopes along with patches of aspen. The drier, south-facing slopes primarily support aspen, sagebrush and gambol oak. Clusters of cottonwood, willow and alder occur along the banks of the Fryingpan River.

Some areas of historical significance near Ruedi Reservoir are Hell Gate and Hagerman Tunnels. Both are associated with the old Midland Railroad, which connected Basalt and Leadville. Coke ovens located in Sellar Park were used to coke coal for the operation of the engines. Evidence of past settlement in the valley include the limestone kilns at Thomasville and the Norrie Townsite.

Wildlife known to inhabit the reservoir include elk, mule deer, and black bear. Small game and non-game species include blue grouse, snowshoe hare, cottontail, porcupine, marmot, weasel, marten and coyote. Common birds include mallards, red-tailed hawks, magpies, and hummingbirds. Bald eagles are often sighted wintering in the area.

CONTROLLING AGENCY White River National Forest.

INFORMATION Sopris Ranger District (970) 963-2266.

HANDICAPPED ACCESS Some handicapped acces- sible facilities at the Mollie B. Campground.

MAP REFERENCES White River National Forest map. Benchmark's *Colorado Road & Recreation Atlas:* p. 72.

Sanchez Reservoir

To San Luis

242

P

Minimum Pool 8272'

Maximum Pool 8317'

N

Scale
0   1/4   1/2   3/4   1 mile

Smith Reservoir

Fort Garland

To 25

160

Rd. 51

59

Rd. 49

San Luis

142

Rd.

159

Sanchez Reservoir

Rd. 85

Sanchez Reservoir

DIRECTIONS From San Luis 3 miles east on Colorado 152 to Colorado 242 and head 5.2 miles south to reservoir.

FEE No.

SIZE 4,571 acres.

ELEVATION 8,317 feet.

FACILITIES Restrooms.

BOAT RAMP One Boat Ramp.

FISH Northern pike, walleye, perch. Fishing for pike and perch is rated excellent.

RECREATION Fishing. No waterskiing. Public access prohibited except for fishing.

CAMPING Dispersed. Camping prohibited in the boat ramp parking area.

CONTROLLING AGENCY Colorado Division of Wildlife.

INFORMATION Monte Vista Office (719) 587-6900.

MAP REFERENCES Benchmark's *Colorado Road & Recreation Atlas:* p. 126.

San Luis State Park

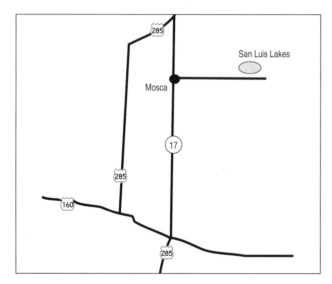

DIRECTIONS From Alamosa, travel north on Colorado 17 for 14 miles to Mosca. Then drive east on 6 Mile Lane (Sand Dunes Highway) about 8 miles to the park entrance.

FEE Daily or annual pass.

SIZE 890 acres.

ELEVATION 7,519 feet.

MAXIMUM DEPTH 10 feet.

FACILITIES Restrooms, bathhouse, laundry, park office, dump station, and drinking water.

BOAT RAMP Concrete ramp near park entrance.

FISH Shallow reservoir rated excellent for rainbow trout. Carp at north end of the lake. Check for size, bag and possession limits.

RECREATION Boating, fishing, wind surfing, waterskiing, picnicking, hiking, biking, fishing, hunting, swimming, bird watching.

CAMPING The Mosca Campground is located in the low sand dunes just west of San Luis Lake. There are 51 campsites. All have electrical hookups, sheltered tables, and fire grates. The campground can accommodate motor homes, trailers and tents. A bathhouse with modern restrooms, hot showers and laundry facilities are located in the camp-

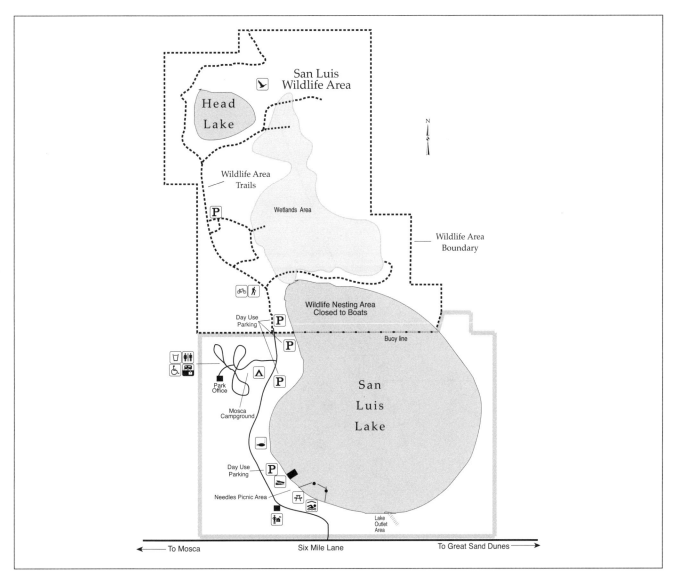

ground. There are no individual sewer hookups, but a dump station is available. All campsites offer panoramic views of the lake and the Sangre de Cristo Mountains. For information about group camping contact the park office.

OTHER INFORMATION The combination of wetlands, lakes and dry valley floor environment provides a fantastic year-round wildlife and recreation area. Migratory waterfowl and birds are frequent visitors to the lakes. Foxes, coyotes, raccoons, pronghorn and small mammals are a common sight in the surrounding low dunes. Songbirds, raptors, reptiles, and amphibians all find refuge in this unlikely riparian oasis.

Nestled in the heart of an alpine valley, in the shadow of the spectacular Sangre de Cristo Mountains, near the edge of the world famous Great Sand Dunes National Monument lies San Luis State Park.

The natural lakes and wetlands of this valley oasis provide excellent fishing and boating in an outstanding setting.

Surrounded by eight of Colorado's awesome fourteeners, the scenic splendor of snow-capped peaks, crystal air and broad vistas delight the eye in all directions.

The wildlife and water supply of the lakes has attracted humans to the area for tens of thousands of years as the artifacts of Folsom Man found nearby indicate. Pueblo Indians believe their people originated under the sparkling water of San Luis Lake.

CONTROLLING AGENCY Colorado State Parks (719) 378-2020.

HANDICAPPED ACCESS Handicapped accessible campsites, showers, picnic tables, fishing access and restrooms. Some facilities accessible with assistance.

MAP REFERENCES Benchmark's *Colorado Road & Recreation Atlas:* p. 114.

Smith Reservoir

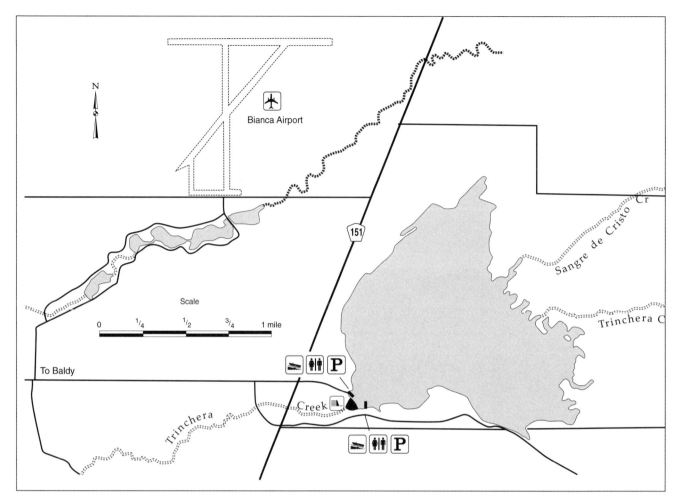

DIRECTIONS From Blanca drive 0.5 miles west on US 160 to Airport Road. Then travel four miles south to the reservoir.

FEE No.

SIZE 700 acres.

ELEVATION 7,721 feet.

FACILITIES Restrooms. No drinking water.

BOAT RAMP Two boat ramps.

FISH Trout and channel catfish. Fishing for rainbow is rated good. Fishing is prohibited on northern portion of the reservoir between November 1 to the last day of waterfowl season. Public access is prohibited February 15–July 15 on north and east shores.

RECREATION Fishing, boating, hunting. No waterskiing allowed.

CAMPING Dispersed camping.

CONTROLLING AGENCY Colorado Division of Wildlife.

INFORMATION Monte Vista Area Office (719) 587-6900.

MAP REFERENCES Benchmark's *Colorado Road & Recreation Atlas:* p. 126.

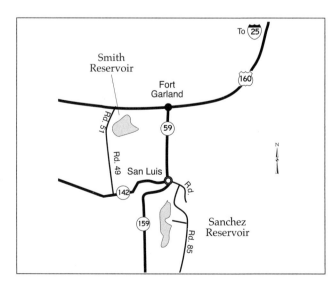

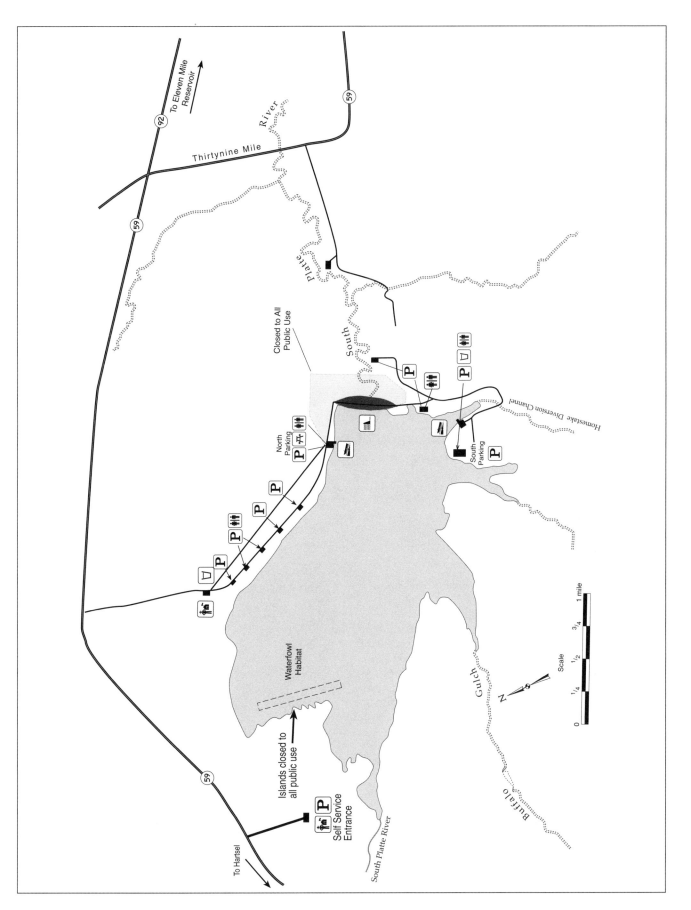

To Eleven Mile Reservoir

92

River

Thirtynine Mile

59

59

Platte

South

Closed to All Public Use

Homestake Diversion Channel

North Parking

P

P

South Parking

P

Waterfowl Habitat

Islands closed to all public use

Gulch

Buffalo

Scale

N

0   1/4   1/2   3/4   1 mile

P

Self Service Entrance

South Platte River

To Hartsel

Spinney Mountain State Park

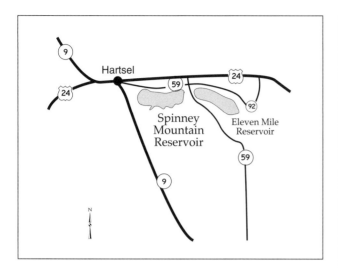

DIRECTIONS From Denver, travel south on I-25 to Colorado Springs. Travel west 55 miles from Colorado Springs on Colorado 24, over Wilkerson Pass. Turn left on County Road 23 and travel 2.8 miles, then turn right on County Road 59. Travel 1.1 miles to the park entrance.

FEE Daily or annual pass.

SIZE 2,520 acres.

ELEVATION 8,687 feet.

MAXIMUM DEPTH 55 feet.

FACILITIES Restrooms, fire rings, picnic tables, water hydrant.

BOAT RAMP One 3-lane concrete ramp at the north picnic area and a 2-lane ramp at the south picnic area.

FISH The reservoir's main attraction is renowned trophy fishing. The area is for day-use only, and its season lasts from approximately mid–April through mid–November.

Cutthroat, rainbow and brown trout and northern pike. Because Spinney is Gold Medal water, only artificial flies and lures may be used. Suggested spoons: Tasmanian Devils, Thomas Buoyants and Cyclones, Kastmasters, Dick Nites, Krocodiles and Super Dupers. Jigs: Gitzits and Lead Head Curly Tails. Spinner flies: Woolly Worms, Woolly Buggers and Scuds. Flies: Woolly Worms, Nymphs, Shrimp patterns and large egg patterns. Early fishing at Spinney is best in shallow water, mainly because the shallow water areas will be the warmest. Generally, the north shoreline will get warmer first because of the southern exposure and wind protection. Fishermen may keep only one trout, which must be twenty inches or longer. Check for current bag and possession regulations.

RECREATION Fishing, boating, hiking, wind surfing, hunting, picnicking, wildlife observation.

CAMPING No overnight camping is allowed at Spinney. A 350-site campground is available at Eleven Mile State Park, 7 miles east of Spinney. A camping fee is charged and park passes are required. Dump stations, picnic sites, restrooms, laundry, showers and fire grates are available.

CONTROLLING AGENCY Colorado State Parks.

INFORMATION Lake George Office (719) 748-3401.

RESTRICTIONS
1. Ice fishing prohibited; reservoir closed to public use from approximately Mid-November through Mid-April.
2. Fishing and boating are prohibited from a half-hour after sunset to a half-hour before sunrise.
3. Fishing is prohibited immediately below the dam.
4. Overnight camping is prohibited.
5. Open fires are permitted only in the grate facilities in the picnic areas.
6. Boat launching facilities are provided at the north picnic area and south picnic area.
7. Vehicle use is restricted to designated roads only.
8. Pets must be leashed.
9. Colorado Parks Pass required to enter.
10. Restrictive fishing regulations in force. Check Division of Wildlife regulation brochure.
11. Waterskiing and swimming are prohibited.

OTHER INFORMATION A high plateau reservoir in South Park, Spinney is set in an open, windswept region. It's shadeless, open and known for constant weather changes during the course of the day.

HANDICAPPED ACCESS Handicapped accessible restrooms, picnic tables and fishing. Some facilities accessible with assistance.

MAP REFERENCES Benchmark's *Colorado Road & Recreation Atlas:* p. 88.

Spring Creek Reservoir

ELEVATION 9,915 feet.

FACILITIES Vault toilet, tables, firewood, drinking water, and trash service.

BOAT RAMP Non-motorized, hand-powered only boat ramp.

FISH Rainbow, cutthroat and brown trout. Good action for shore fishermen using bait or fly fishing. The best fishing typically has been found in the early morning or late afternoon.

RECREATION Canoeing, fishing.

CAMPING Mosca Campground offers 16 sites with access for RVs up to 35 feet in a beautiful mountain setting. Reservations are not accepted.

CONTROLLING AGENCY Grand Mesa, Uncompahgre, and Gunnison National Forests.

INFORMATION Gunnison Ranger District (970) 641-0471.

MAP REFERENCES Benchmark's *Colorado Road & Recreation Atlas:* p. 86.

Spring Creek Reservoir

744

Mosca Campground

N

744

F.S. Rd.

28 miles to Gunnison

748

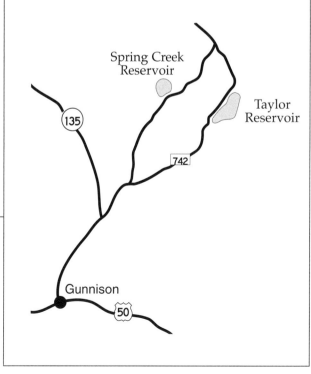

DIRECTIONS From Gunnison drive north on Colorado 135 to the town of Almont. From Almont travel 7 miles northeast on County Road 742 to the vicinity of Harmel's Resort. Then drive north 10 miles on Forest Road 744 following alongside Spring Creek to reservoir.

FEE Camping fees.

SIZE 50 surface acres.

St. Vrain State Park

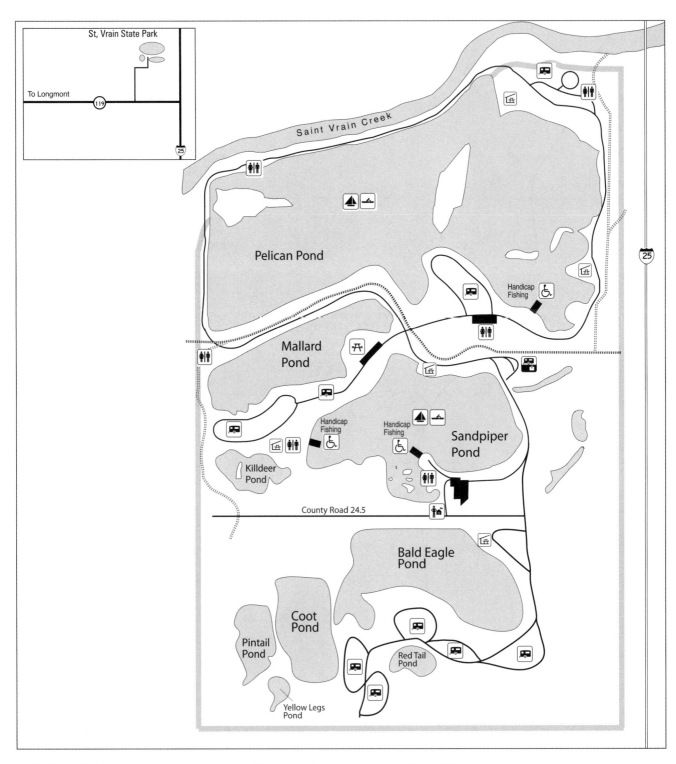

DIRECTIONS Drive north from Denver on I-25 to the Colorado 119 exit to Longmont. Travel west on Colorado 119 approximately 2 miles to the park entrance on the north side of the highway.

FEE Daily or annual pass.

SIZE 236 acres (not all available to public).

ELEVATION 4,830 feet.

MAXIMUM DEPTH 6 feet (1974 survey).

FACILITIES Dump station, picnic tables, fishing piers and platforms, and restrooms.

BOAT RAMP None, small sailing and hand-propelled craft

only, includes air-inflated devices if more than one compartment.

FISH The ponds contain bluegill, yellow perch, carp, bass, channel catfish, crappie and rainbow trout. Rainbow trout are stocked every spring and fall. Beginning in late March, the stocking program provides great fishing. Fall stocking begins in late September for fall and ice fishing.

RECREATION Boating, fishing, hiking, icefishing, camping. Swimming is prohibited.

CAMPING In July 2007, the phase II area opened in the park. There are now a total of 87 campsites open for camping. Sites 1-41 have electric hookups with water hydrants available to fill tanks and a dump station at the exit of the campground. Sites 42-87 have water, sewer, and electric hookups. Phase II also added several new fishing ponds and two additional hiking trails. Further development within the park is planned.

OTHER INFORMATION The group of ponds in St. Vrain State Park, which lie in the shadow of Longs Peak, can be found just 30 miles north of Denver off I-25. St. Vrain State Park began with a group of drab gravel pits that had been dredged for highway construction. A reclamation project began in 1962 when the Colorado Game, Fish and Parks Department obtained the area from the Highway Department.

The park is full of recreational opportunity, both during and after the fishing season. The ponds are well stocked with rainbow trout and catfish offering some of the best warm water fishing in the northern part of the state. Ice fishing is excellent. The area is closed to all types of hunting.

Geese and many species of ducks inhabit the area and visitors will see great blue herons and may even enjoy an occasional glimpse of a black-crowned night heron. There is an abundance of other wildlife and plants as well.

The shaded pond-side picnic area has six sites and is popular with day visitors. Swimming is prohibited, but visitors can sail or use rowboats or inflatables with multiple compartments. There are special requirements on certain boats and inflatables. Check with the park ranger to be sure yours is acceptable.

Snacks and fuel are available on Colorado 119, two miles from the park.

HANDICAPPED ACCESS The area around the ponds is relatively flat, making St. Vrain State Park largely accessible for the physically challenged, especially with some assistance. Concrete ramps are provided to access five of the six toilets in the park. Campsite #45 is fully accessible. There are four fully accessible parking stalls at each jetty. Some facilities available with assistance.

CONTROLLING AGENCY Colorado State Parks.

INFORMATION Firestone Office (303) 678-9402.

MAP REFERENCES Benchmark's *Colorado Road & Recreation Atlas:* p. 63.

Stagecoach State Park

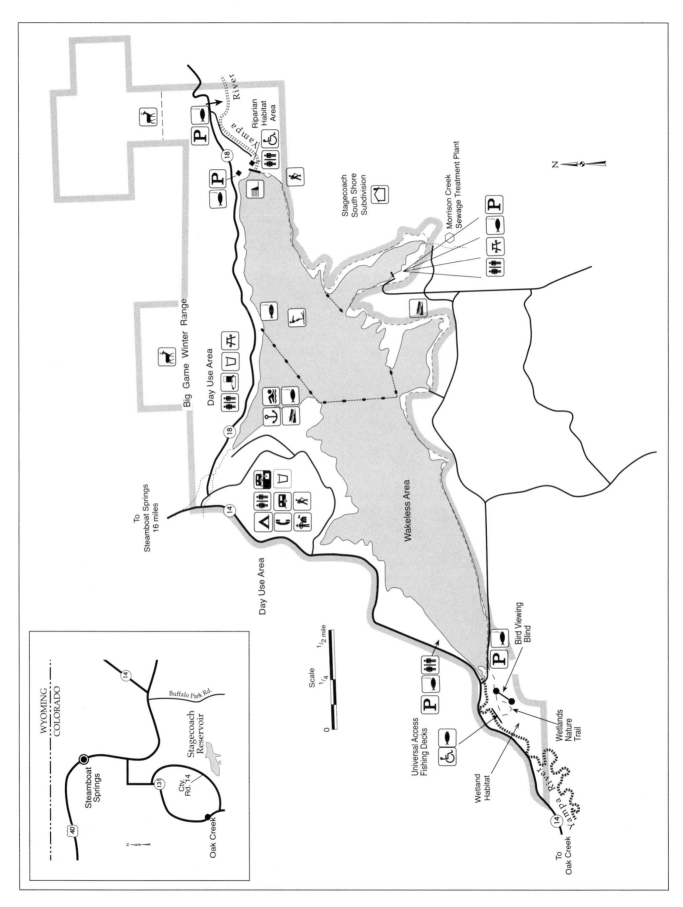

Stagecoach State Park

DIRECTIONS From Denver, take I-70 west to the Silverthorne exit, Colorado 9. Travel north on Colorado 9 to Kremmling. North of Kremmling, travel west on Colorado 134 across Gore Pass to Toponas. Take Colorado 131 north out of Toponas through Phippsburg. Turn right on County Road 14 at the Stagecoach Reservoir State Recreation Area sign.

FEE Daily or annual pass.

SIZE 775 acres.

ELEVATION 7,210 feet.

MAXIMUM DEPTH 145 feet.

FACILITIES Swim beach, park office, dump station, boat rentals, picnic areas, and marina offering rentals, a gas dock, and basic fishing and camping supplies.

BOAT RAMP A boat ramp is located at the marina near Pinnacle Campground and at the Morrison Cove area. The reservoir is divided into wakeless and non-wakeless areas.

FISH Northern pike, kokanee salmon, Tasmanian rainbows, cutthroats, brook, and brown trout. For trolling, use Needlefish, Kastmaster, Panther Martins, Mepps and Dick Nite. Lures in rainbow patterns work at depths of 17 to 23 feet. Also good is Pop Gear with nightcrawlers weighted to sink 15 to 20 feet. The key is to fish below the 12 to 14 thermal cline. For bank fishing, the best areas are at the north shore of the park around the Keystone picnic area or on County Road 18 below the self-service area. Best baits are Berkley Power Bait, marshmallows and inflated nightcrawlers fished just off the bottom.

RECREATION Fishing, waterskiing, swimming, personal water craft, boating, biking, hiking, horseback riding, and snowmobiling.

CAMPING Four campgrounds have 92 campsites combinrd. All have drinking water, flush toilets and dump site facilities. Electrical hookups are available at 65 sites. A camper service area, showers and pull-through campsites are also available. McKindley Campground offers group accommodation for up to 45 people.

OTHER INFORMATION Wagon wheels and ore cars are two symbols of the area's rich history. Stagecoaches traveled over Yellow Jacket Pass, now County Road 14, on their way to Oak Creek or Steamboat Springs. The Stagecoach site is part of what locals term Egeria Park, a large open area bounded by Oak Ridge to the west and Green Ridge to the east. The name Egeria is Ute for "Crooked Woman," depicting the winding course of the Yampa River through the valley. The Ute Indians frequently passed through the area enroute to their northern territory.

A mail and stage route was located over Lynx Pass. A stage stop was also located near the main park entrance across from the Henderson Ranch.

The three-mile-long Stagecoach Reservoir opened in July 1989 and offers many recreational opportunities. The reservoir, which feeds off the Yampa River, has four major coves. The reservoir sits below black timber and aspen groves.

Mild summer temperatures make for pleasant camping, picnicking, fishing, boating, and swimming. High snow falls and consistently low temperatures provide excellent conditions for winter activities such as ice fishing and cross-country skiing. Facilities include 92 campground units, picnic areas, swim beach, marina, showers, flush toilets and a five mile biking and hiking trail.

SPECIAL RESTRICTIONS Blacktail Mountain provides winter range for elk. The public is prohibited from using the wintering range from January to March. Hunting season is subject to regulations from the first Tuesday after Labor Day until the last day of February.

CONTROLLING AGENCY Colorado State Parks.

INFORMATION Oak Creek Office (970) 736-2436.

HANDICAPPED ACCESS Handicapped accessible showers, restrooms, picnic tables, camping and fishing. Some facilities accessible with assistance.

MAP REFERENCES Benchmark's *Colorado Road & Recreation Atlas:* p. 60.

Steamboat Lake State Park

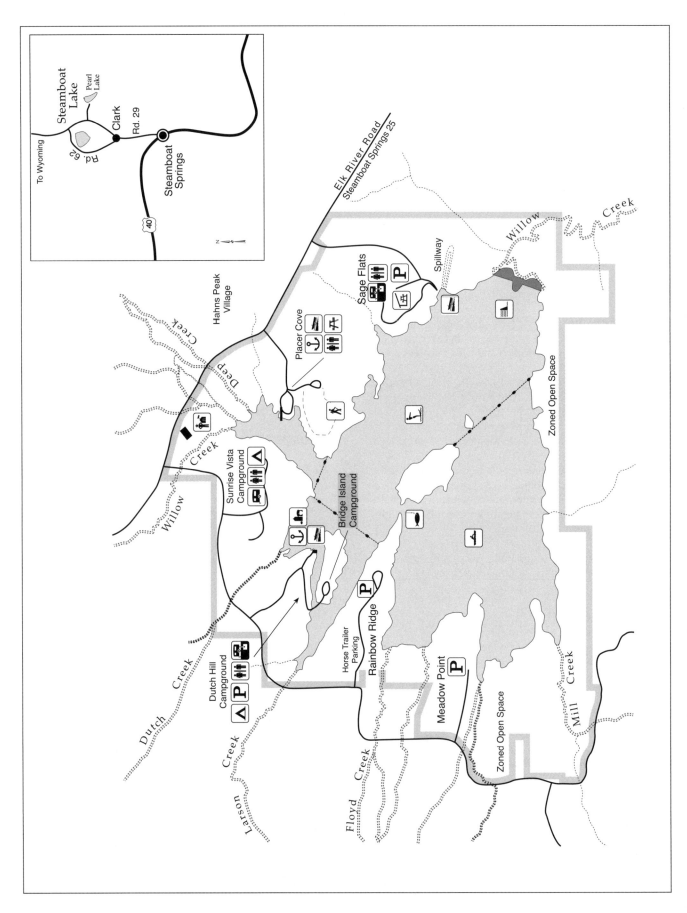

Steamboat Lake

Pearl Lake

Clark

Rd. 29

To Wyoming

Rd. 62

Steamboat Springs

40

Elk River Road
Steamboat Springs 25

Willow Creek

Spillway

Sage Flats

Placer Cove

Hahns Peak Village

Deep Creek

Zoned Open Space

Sunrise Vista Campground

Willow Creek

Bridge Island Campground

Dutch Creek

Horse Trailer Parking

Rainbow Ridge

Dutch Hill Campground

Larson Creek

Floyd Creek

Meadow Point

Zoned Open Space

Mill Creek

Steamboat Lake State Park

DIRECTIONS   From Steamboat Springs drive north on County Road 129 (Elk River Road) to the reservoir.

FEE   Park pass and camping fee.

SIZE   1,058 surface acres (Pearl Lake—190 surface acres).

ELEVATION   8,000 feet.

FACILITIES   Full service state park with a dump station, as well as marina with boat rentals.

BOAT RAMP   Three ramps are located on the north side of reservoir at Sage Flats, Placer Cove and the marina.

FISH   Rainbow, brown and cutthroat trout. Shore or boat fishing in the coves or at the inlets produce 10- to 17-inch trout. Pearl Lake contains cutthroat trout, and artificial lures only are to be used. There is a two fish limit of 18 inches or greater at Pearl Lake.

RECREATION   All water sports, fishing, hiking, hunting, swimming, picnicking, cross-country skiing, ice fishing, snowmobiling.

CAMPING   Sunrise Vista, Dutch Hill and Pearl Lake campgrounds have a total of 224 campsites, which can accommodate tents and campers or trailers. There are some pull-through sites in each campground. Only one unit per site is allowed. ten cabins and two yurts are available for rent.

OTHER INFORMATION   Steamboat Lake, in the northernmost part of Colorado, is nestled in a valley at the foot of majestic Hahns Peak only a few miles west of the Continental Divide. It forms the major portion of Steamboat Lake State Park, which also includes Pearl Lake. Steamboat Lake is a man-made lake that was completed in 1967 and filled in 1968. Pearl Lake, quite small compared to Steamboat, was completed in 1962 and in 1966 was named in honor of Mrs. Pearl Hartt, who was instrumental in aiding the state's acquisition of the land for the construction of the lake. A local resident, Mrs. Hartt was one of the many in the community who helped to make Steamboat Lake Park a reality. Steamboat Lake received its name because of its proximity to nationally known Steamboat Springs.

Because of its location in a valley, the area offers one of the most beautiful settings in Colorado, no matter what the season. Steamboat Lake State Park is famed for water sports, fishing and winter sports and especially as a place to simply relax. It lies adjacent to Medicine Bow & Routt National Forests, providing convenient trail connections for hikers. Also, it is only 27 miles from the resort town of Steamboat Springs which offers bright lights and opportunity for those who desire a "night on the town."

Anglers will be pleased to note that coves at Steamboat Lake are well known for good-sized trout. Pearl Lake is known for cutthroat. Only fly and lure fishing is permitted here. Placer Cove is the spot for ice fishing.

Steamboat Lake is the setting for water sports of all kinds. With the exception of the coves, all of Steamboat Lake is open to waterskiing. Boats towing skiers must keep 150 feet from the shore and swim areas. Boaters are subject to current Colorado boating statutes and regulations. Ski pattern on Steamboat Lake is counterclockwise. Except during legal hunting season, the use of firearms is prohibited in the park.

Many kinds of waterfowl nest in the park and marshes around it. These include mallards, teal, geese, mergansers and cranes. Elk, deer, bear, coyote, fox, chipmunks and other small mammals are found in the area. There is a half-mile nature trail and an amphitheater that hosts interpretive programs during warm weather.

Besides ice fishing, winter sports enthusiasts find the park a good place for snowmobiling and cros-country skiing.

Groceries, phone and boat rentals are available at the park concession. At Clark, seven miles south, there is a post office and groceries, phone, gas and licenses. Minor first-aid is offered by park rangers; first-aid is also available at the information center when open. In an emergency, an ambulance can be summoned from the Clark Fire Department. Routt Memorial Hospital is located in Steamboat Springs, 30 minutes from the park.

CONTROLLING AGENCY   Colorado State Parks.

INFORMATION   Clark Office (970) 879-3922.

HANDICAPPED ACCESS
Steamboat: Visitor center, restrooms, swimming, hunting, picnic tables, camping and fishing.
Pearl: Restrooms, picnic tables, camping and fishing. Some facilities accessible with assistance.

MAP REFERENCES   Benchmark's *Colorado Road & Recreation Atlas:* p. 45.

Strontia Springs Reservoir

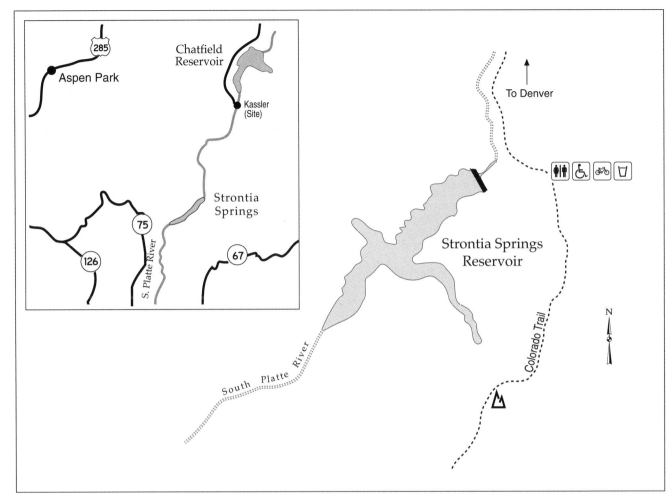

DIRECTIONS From Denver, travel south on Colorado 75 (west of Chatfield State Park) to Kassler. Hike 6 miles southwest on the old road along the South Platte River. The access is relatively easy to moderate for the first couple of miles and somewhat more difficult after that. Three rest areas provide restrooms and benches. From the South Platte River end, heading north up the canyon, it's about a 2 mile hike to the reservoir.

FEE No.

SIZE 98 surface acres.

ELEVATION 5,990 feet.

MAXIMUM DEPTH 212 feet.

BOAT RAMP No ramps; boating is not allowed.

FISH Rainbow trout (catchable-size stocked), brown trout and suckers. Fair to good action on typical baits such as salmon eggs and Power Bait. Suggested lures are the Blue Fox, Mepps and Roostertails in silver on cloudy days and gold on sunny days. Fly patterns that have produced are the Blue Wing Olive, Blue Duns and assorted nymph patterns.

A special bag limit of two fish is in effect from the Scraggy View Picnic Grounds on the South Platte, downstream to the Strontia Springs Dam.

RECREATION Hiking, biking, horseback riding, and wildlife viewing.

CAMPING No camping available in the Waterton Canyon area. Access is walk-in, day-use only on the old road bed. Steep canyon walls surround both sides of the South Platte River. The most accessible camping can be found in the US Forest Service areas along the South Platte River, upstream or south of the canyon.

OTHER INFORMATION Located in Waterton Canyon, the old narrow gauge railroad grade on the north end of the canyon provides 6 miles of relatively good access by foot or bicycle into the canyon along the South Platte River.

CONTROLLING AGENCY Denver Water Department.

INFORMATION Colorado Division of Wildlife (303) 291-7227.

MAP REFERENCES Pike National Forest map. Benchmark's *Colorado Road & Recreation Atlas:* p. 75.

Sweetwater Lake

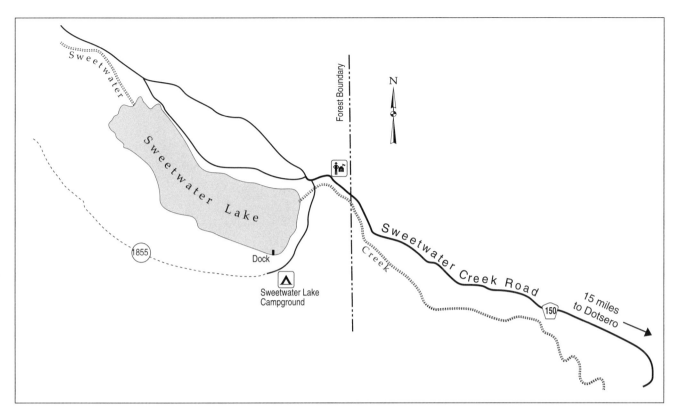

DIRECTIONS From Denver, travel west on Interstate 70. Exit at the town of Dotsero (Dotsero is 72 miles west of Dillon) and follow the signs for Sweetwater. Turn north onto the Colorado River Road and travel 7 miles to the Sweetwater Creek Road. Turn left onto Sweetwater Creek Road and travel 10 miles, until you see a sign for Sweetwater Campground. Turn left at the sign and travel to Sweetwater Lake.

FEE Camping fee.

SIZE 72 surface acres.

ELEVATION 7,709 feet.

FACILITIES Pit toilets.

BOAT RAMP There are no public ramps, and boating is restricted to hand-launch, electric or hand-powered craft. Gasoline-burning engines are not allowed. During the summer there is a public boat dock located near the campground.

RECREATION Horseback riding, fishing, hiking, spelunking.

CAMPING Sweetwater Campground offers 10 sites, 6 with RV access 30 feet or less. Picnic tables, fire ring, grill and trash disposal. The surrounding area is relatively dry with scrub oak, douglas fir and serviceberry. High limestone cliffs flank the lake on the east and west sides.

FISH Brook, brown and rainbow trout and kokanee salmon. Fishing in this mostly local fishing hole is rated good using typical baits from shore. The best action from small boats is trolling some type of flasher tipped with a nightcrawler. Fair to good action fly fishing or lure fishing in the early mornings or evenings. Mostly stocker-sized rainbow trout are taken. Sweetwater Creek runs into and out of the lake. The water levels tend to remain the same.

CONTROLLING AGENCY White River National Forest.

INFORMATION Eagle Ranger District: (970) 328-6388.

MAP REFERENCES White River National Forest map. Benchmark's *Colorado Road & Recreation Atlas:* p. 71.

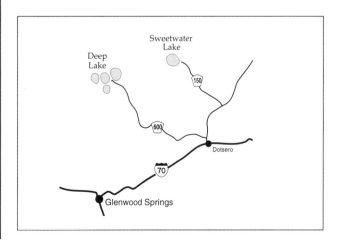

Sylvan Lake State Park

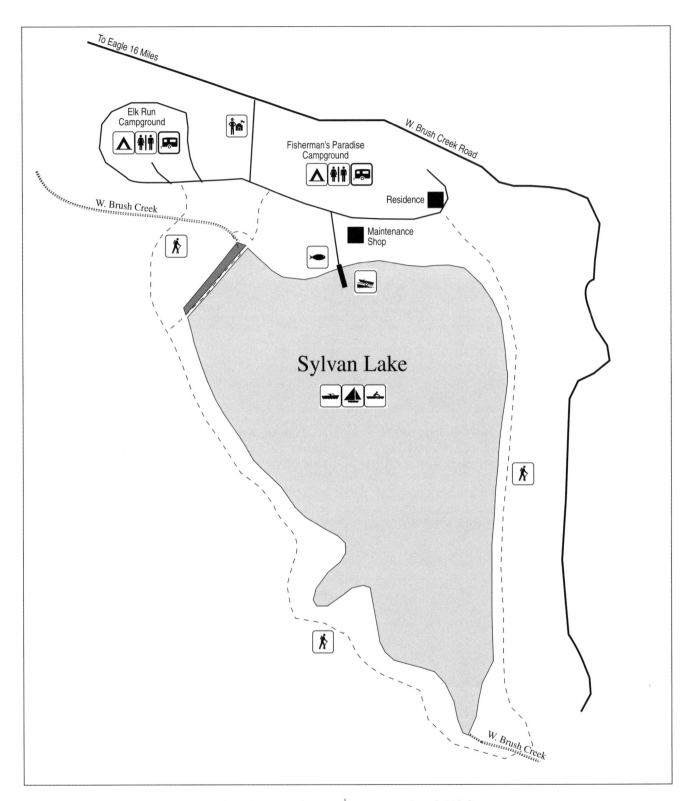

DIRECTIONS Travel 16 miles south of Eagle on Brush Creek Road to Sylvan Lake.

FEE Daily or annual pass.

SIZE 42 acres.

ELEVATION 8,510 feet.

FACILITIES Water, restrooms, showers, visitor center, trails, group campground, tables, grills.

BOAT RAMP One ramp located near Fisherman's Paradise

Sylvan Lake State Park

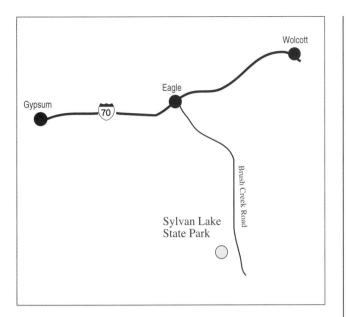

CAMPING Sylvan Lake has 44 campsites, 32 at Elk Run Campground and 12 at Fisherman's Paradise Campground. A camper services building provides hot showers and flush toilets. Sites can accommodate tents, trailers and campers. There are some pull-through sites for larger units. Campers must have camping permit. Nine cabins and three yurts are available for rent.

OTHER INFORMATION Sylvan Lake State Park is one of the most beautiful getaways in the state. Nestled in the heart of the majestic Rocky Mountains, it is surrounded by the White River National Forest. The visitor enjoys a spectacular 360-degree panoramic view of the alpine scenery—one of Colorado's best kept secrets.

Not only is the park one of the most scenic getaways, it has some of the best year-round trout fishing anywhere.

Sylvan Lake is a haven for campers, boaters, picnickers, photographers and hikers. The park is used as a base camp for hunters during the big game hunting season.

CONTROLLING AGENCY Colorado State Parks.

INFORMATION Eagle Office (970) 625-1607.

MAP REFERENCES Free park brochure.
Benchmark's *Colorado Road & Recreation Atlas:* p. 72.

Campground. Only non-motorized boats or boats with electric motors allowed.

RECREATION Boating, fishing, hiking, biking, horseback riding, hunting, cross-country skiing, snowmobiling.

FISHING Stocked with brook, brown, cutthroat, and rainbow trout.

Tarryall Reservoir

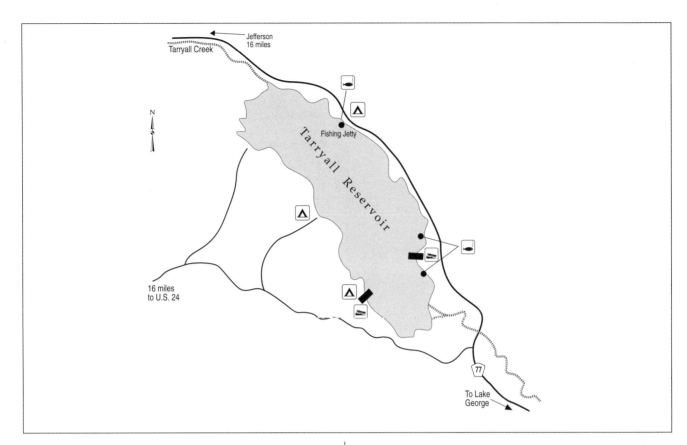

DIRECTIONS   From Denver, take US 285 south to Jefferson. Drive southeast for 15 miles on County Road 77 to reservoir. From Lake George: Go one mile west on US 24 to Tarryall Road then go northwest 22 miles to reservoir.

FEE   Habitat stamp required.

SIZE   165 acres.

ELEVATION   8,860 feet.

FACILITIES   Restrooms, boat ramps, fishing piers.

BOAT RAMP   Two gravel ramps are available. One ramp is on the northeast shoreline, the other is by the campground on the southwest side.

FISH   Rainbow, brook and brown trout. Good results from boats trolling flashers and Pop Gear with half a nightcrawler trailing. Shore fishing has produced good results using Power Bait, salmon eggs and worms on the bottom. Mepps, Rooster Tail and Panther Martin lures also have been effective.

RECREATION   Fishing, boating, hunting, wildlife viewing.

CAMPING   Open dispersed camping in the state wildlife area. Vault toilets are available, but no other facilities are offered.

OTHER INFORMATION   Located in South Park, this huge intermountain valley is surrounded by the Continental Divide on the west and Pikes Peak on the east.

CONTROLLING AGENCY   Colorado Division of Wildlife.

INFORMATION   Denver Office (303) 291-7227.

MAP REFERENCES   Pike National Forest map. Benchmark's *Colorado Road & Recreation Atlas:* p. 88.

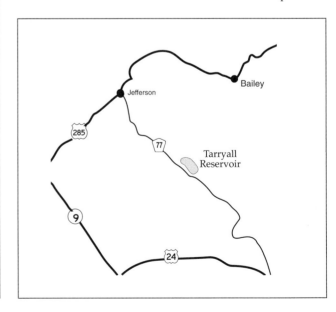

Taylor Park Reservoir

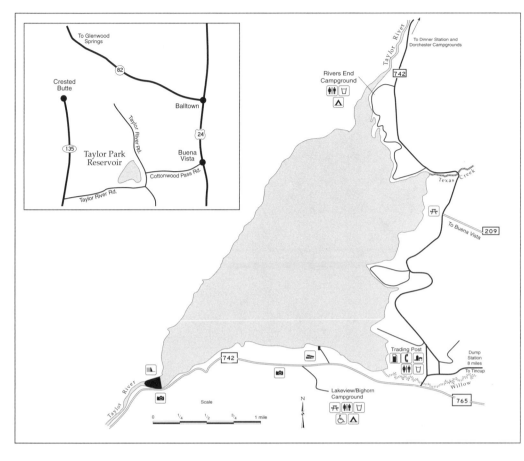

lor River features 17 miles of river fishing. Watch for private land. Some catch and release waters.

RECREATION Fishing, boating.

CAMPING Rivers End Campground on the north end of the reservoir and Lakeview Campground on the south end are US Forest Service campgrounds. Rivers End has 15 campsites, and Lake View has 35 sites. Both campgrounds offer vault toilets, pump water and trash disposal. Reservations accepted at Lakeview Campground. Seven Forest Service campgrounds are available along the Taylor River.

Taylor Reservoir is flanked by great mountain scenery and a high-country forest.

DIRECTIONS From Pueblo, travel west on US 50 to Gunnison, then north from Gunnison on Colorado 135 to Almont. From Almont, travel northeast along the Taylor River on Forest Road 742 to the Taylor Park Reservoir.

FEE Camping fee.

SIZE 2,400 surface acres.

ELEVATION 9,330 feet.

MAXIMUM DEPTH 130 feet.

FACILITIES Restrooms, drinking water, marina with boat rentals.

BOAT RAMP Suitable launch site at boathouse on south side of reservoir.

FISH Rainbow, brown and cutthroat trout, sucker, mackinaw. kokanee and northern pike. Fishing is good off the bank for cutthroat, rainbow and browns on salmon eggs, Power Bait or worms. The best result from boats are obtained by jigging brown and orange or green and orange lures along the northwest shoreline. Good action has been reported for northern pike on a local pattern jig called the Fry Bug. Mackinaw can be caught on chartreuse, gitzits and sucker meat. The drive to the reservoir along the Tay-

CONTROLLING AGENCY Grand Mesa, Uncompahgre, and Gunnison National Forests.

INFORMATION Gunnison Ranger District (970) 641-0471.

MAP REFERENCES Benchmark's *Colorado Road & Recreation Atlas:* p. 86.

Trinidad Lake State Park

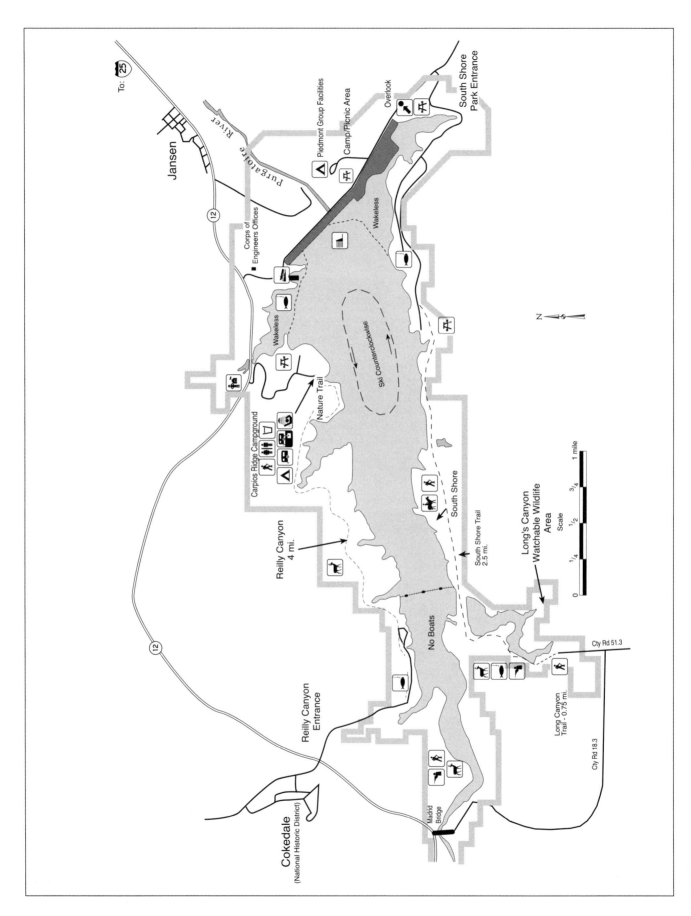

To: 25

Jansen

Purgatoire River

12

Corps of
Engineers Offices

Piedmont Group Facilities

Camp/Picnic Area

Overlook

South Shore
Park Entrance

Wakeless

Wakeless

Ski Counterclockwise

Nature Trail

Carpios Ridge Campground

Reilly Canyon
4 mi.

South Shore

South Shore Trail
2.5 mi.

Long's Canyon
Watchable Wildlife
Area

Scale

0   1/4   1/2   3/4   1 mile

Cty Rd 51.3

Long Canyon
Trail - 0.75 mi.

Cty Rd 18.3

No Boats

Reilly Canyon
Entrance

Madrid
Bridge

12

Cokedale
(National Historic District)

N

Trinidad Lake State Park

DIRECTIONS Travel three miles west from Trinidad on Colorado 12 to the reservoir.

FEE Daily or annual pass.

SIZE 800 acres.

ELEVATION 6,225 feet.

MAXIMUM DEPTH 200 feet.

FACILITIES Full service state park with a dump station.

BOAT RAMP One ramp located on the north side near the dam. There are also seasonal docks.

FISH Rainbow and brown trout, largemouth bass, channel catfish, walleye, crappie, bluegill and wipers. Trinidad Lake is rated good for saugeye and trout fishing. Check current regulations for size and bag limits. The water level can vary, so be alert for submerged hazards. Fishing is permitted anywhere on the lake except in the boat launching and docking area. There is a boat ramp on the north end of the lake for access to boaters, fishermen and water skiers. All boats must observe wakeless speeds around the boat launch area, within 150 feet of shore fishermen, and as buoyed. Boaters are warned to be especially alert to submerged hazards. As the water level drops, land outcroppings appear, especially just west of the launching area and along the south shore. There are no beaches and swimming is not permitted.

RECREATION Boating, fishing, biking, hiking, horsesback riding, personal water craft, hunting, no swimming.

CAMPING The 62 available sites can accommodate recreation vehicles, trailers, or tents. Modern facilities include water hydrants, coin operated laundry, electrical hookups, showers and flush toilets. Campsites can be reserved in advanced. Campers can stay a maximum of 14 days in any 45 day period. A holding tank dump station is located near the campground. Picnicking is permitted throughout the park. Group picnic and camping facilities are available by reservation at (719) 846-6951.

HUNTING Hunting is not permitted in the park between Memorial Day and Labor Day, but at other times hunting is permitted in specially posted areas during legal seasons. Shotguns or bow and arrow are the only weapons permitted.

HIKING There are hiking trails for those so inclined. Two short trails lead from a trailhead located in the campground/picnic area. Neither has waste or restroom facilities. The three-quarter-mile Carpios Ridge Trail is steep but offers views of the reservoir and Fishers Peak. The Levsa Canyon Trail loops to the west above the reservoir for equally striking views. The hiking enthusiast can continue 4 miles to the Reilly Canyon entrance and the historic town of Cokedale. Across the lake the 2-mile South Shore Trail takes hikers to Longs Canyon and seldom explored areas of the park.

OTHER INFORMATION Trinidad Lake is close to the state line in southern Colorado and lies in the Purgatories River valley. It was originally built as an irrigation and flood control project by the US Corps of Engineers. The area around the lake, now known as Trinidad Lake State Park, has been managed by the Colorado Division of Parks and Outdoor Recreation under a lease agreement since 1980. This 2,700-acre park, with its 800-acre lake, serves as a destination area or as a camp from which to explore points of scenic and historic interest in southern Colorado. It is an area of unique beauty and has been a stopping point for travelers since the early Indian days.

Archaeological finds in and around the park are reminders of that early history. These include tepee rings in the Carpios Ridge Picnic Area and the mountain branch of the Santa Fe Trail that is now one of Trinidad's main thoroughfares.

Interpretive campfire programs are offered every Friday, Saturday, and holiday evenings Memorial Day through Labor Day.

CONTROLLING AGENCY Colorado State Parks and US Corps of Engineers.

INFORMATION Trinidad Office (719) 846-6951.

HANDICAPPED ACCESS The Carpios Ridge Campground and the picnic area have reserved parking spaces and campsites adapted for handicapped persons. Restrooms, showers, picnic tables, trails, camping, fishing and drinking fountains are also accessible. Some facilities accessible with assistance.

MAP REFERENCES Benchmark's *Colorado Road & Recreation Atlas*: p. 128.

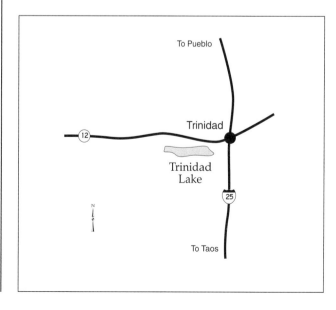

Trujillo Meadows Reservoir

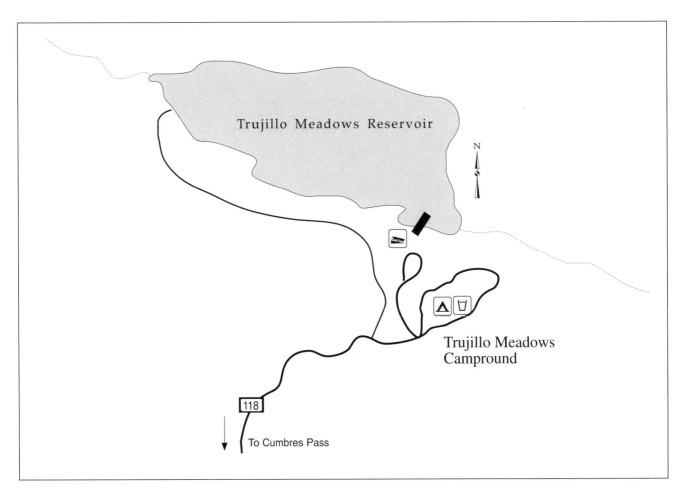

Trujillo Meadows Reservoir

Trujillo Meadows Campround

118

To Cumbres Pass

DIRECTIONS From Antonito, go 36 miles west on Colorado 17 to Forest Road 118 near Cumbres Pass, then head four miles north to the reservoir.

FEE Camping fee.

SIZE 69 acres.

ELEVATION 10,020 feet.

FACILITIES Campground, boat ramp, restrooms.

BOAT RAMP One ramp; boating that creates a whitewater wake is prohibited.

RECREATION Fishing, hiking, wildlife observation.

CAMPING Fifty campsites that will handle units to 25 feet with fire grates and drinking water. No reservations.

FISH Rated good for brown and rainbow trout. Large brown trout caught in the winter through ice.

CONTROLLING AGENCY Department of Wildlife.

INFORMATION Monte Vista (719) 587-6900.

HANDICAPPED ACCESS Handicapped accessible camping and restrooms. Some facilities accessible with assistance.

MAP REFERENCES Rio Grande National Forest. Benchmark's *Colorado Road & Recreation Atlas:* p. 124.

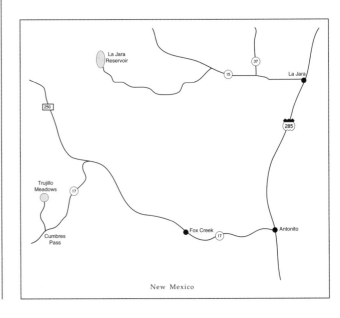

Turquoise Lake

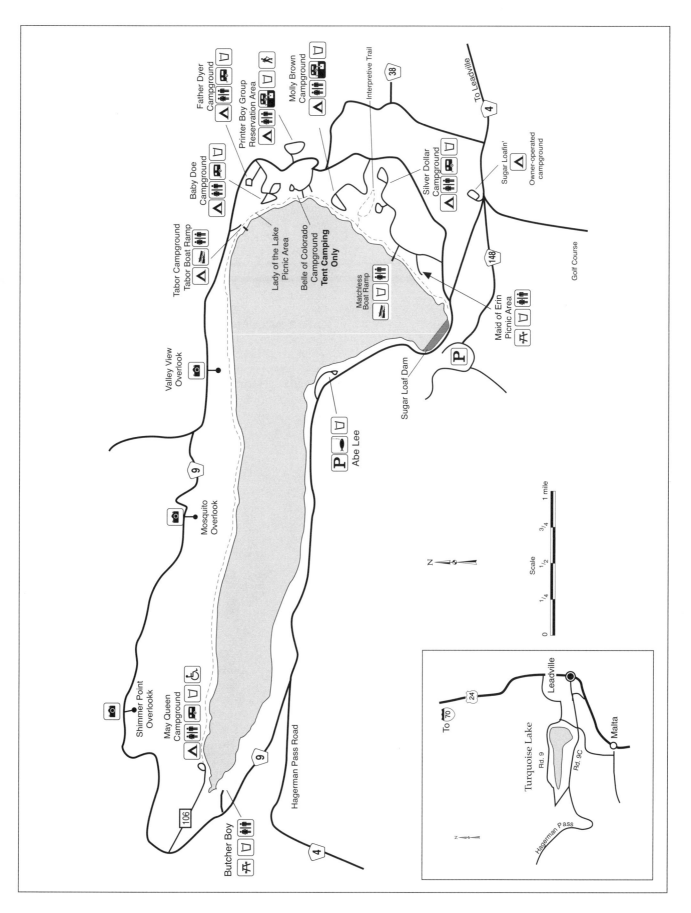

Father Dyer Campground

Printer Boy Group Reservation Area

Molly Brown Campground

Interpretive Trail

38

To Leadville

4

Baby Doe Campground

Silver Dollar Campground

Sugar Loafin'

Owner-operated campground

Tabor Campground
Tabor Boat Ramp

Lady of the Lake Picnic Area

Belle of Colorado Campground
Tent Camping Only

Matchless Boat Ramp

Maid of Erin Picnic Area

148

Golf Course

Valley View Overlook

Abe Lee

Sugar Loaf Dam

P

P

Mosquito Overlook

9

Shimmer Point Overlookk

May Queen Campground

Butcher Boy

106

9

Hagerman Pass Road

4

Scale

N

0   1/4   1/2   3/4   1 mile

To 70

24

Leadville

Malta

Turquoise Lake

Rd. 9

Rd. 9C

Hagerman Pass

N

Turquoise Lake

DIRECTIONS From the intersection of 6th Street and Harrison Avenue in Leadville, drive 0.1 miles west on 6th Street to Mcwethy Drive, then turn right and follow County Road 4 (Turquoise Lake Road) around to the west for 5.5 miles to the lake.

FEE Daily or annual pass. Camping fees.

SIZE 1,788 surface acres when full.

ELEVATION 9,869 feet.

MAXIMUM DEPTH 80 to 90 feet.

FACILITIES Full-service campgrounds.

BOAT RAMP Two concrete boat ramps are available. The Tabor boat ramp is on the northeast end of the lake; the Matchless boat ramp is on the southeast end.

FISH Cutthroat, rainbow, brown and lake trout.

RECREATION Fishing, boating, hiking. Hiking trailheads for Turquoise Lake Interpretative, Turquoise Lake and Timberline Lakes trails.

CAMPING Campgrounds are generally open from Memorial Day weekend until Labor Day weekend. Seven campgrounds are available. On the west end of the lake, the May Queen Campground offers 27 campsites for units up to 32 feet. On the east side of the lake, Baby Doe offers 50 sites for units up to 32 feet, Molly Brown has 49 sites for units up to 32 feet, and Belle of Colorado has 19 sites for tent camping only. Father Dyer has 25 sites for units up to 32 feet, Tabor has 27 sites for units up to 32 feet, and Silver Dollar 43 sites for units up to 22 feet.

Most of the campgrounds have toilets, picnic tables and fire grates. Water also is available. Several picnic areas are located around Turquoise Reservoir. Unreserved campsites are handled on a first-come, first-served basis. A group campsite, Printer Boy, is available. It has a capacity of 220 people. Silver Dollar, Molly Brown, Printer Boy, Baby Doe, and Father Dyer Campgrounds are on a reservation system. Dump stations are located at Baby Doe, Molly Brown and Printer Boy Campgrounds.

OTHER INFORMATION Turquoise Reservoir is located five miles west of Leadville. The lake is man-made and is a part of the Lake Fork drainage in the upper Arkansas Valley. It has an approximate elevation of 9,900 feet in the mountainous terrain and backs against the Sawatch.

Mountain Range, which includes Colorado's two highest peaks, Mount Elbert and Mount Massive.

Turquoise Reservoir is incorporated into the Fryingpan-Arkansas Water Project. This project is diverting water from the west slope of the continental divide to the more populated eastern slope. Water is diverted by a system of tunnels and at Turquoise Reservoir there are two tunnels, the Charles Boustead Tunnel and the Horsestake Tunnel. Turquoise Reservoir is one of several storage areas in the Fryingpan-Arkansas Project.

Leadville, as well as the Turquoise Reservoir area, is rich in a colorful history. Leadville's glamorous past began during the "Gold Rush" in California Gulch, southeast of Leadville, in 1860. Many of Leadville's prominent personalities and places of this era are now honored as recreation sites along Turquoise Reservoir.

In 1860, after leaving the 59ers in the Pikes Peak region, one expedition crossed the Mosquito Range and traveled up the Arkansas River into California Gulch. At this location, one man panned the first valuable pay-dirt. This man was Abe Lee. A fisherman parking area is now honored with his name.

Some five years after the 1860 gold rush into California Gulch, the ore and population started to dwindle. Using new methods of underground "hardrock" mining, one mine began to pay off after several years. The mine was Printer Boy. Now Printer Boy Campground at Turquoise Reservoir carries this name.

In 1877, Alvinus B. Wood and William H. Stevens analyzed some rock and sand which had been clogging prospectors' mining equipment. They found it to be lead carbonate, richly lined with silver. The silver boom was on! Leadville grew from 300 to over 30,000 residents in less than two years. Mining claims and mines sprang up seemingly overnight. Four of these mines are now names of recreation sites in the Turquoise Reservoir Recreation Area. They are Lady of the Lake Picnic Area, Maid of Erin Picnic Area, Belle of Colorado Campground, and May Queen Campground.

CONTROLLING AGENCY Pike & San Isabel National Forests.

INFORMATION Leadville Ranger District (719) 486-0749.

MAP REFERENCES San Isabel National Forest. Benchmark's *Colorado Road & Recreation Atlas:* p. 86.

Twin Lakes

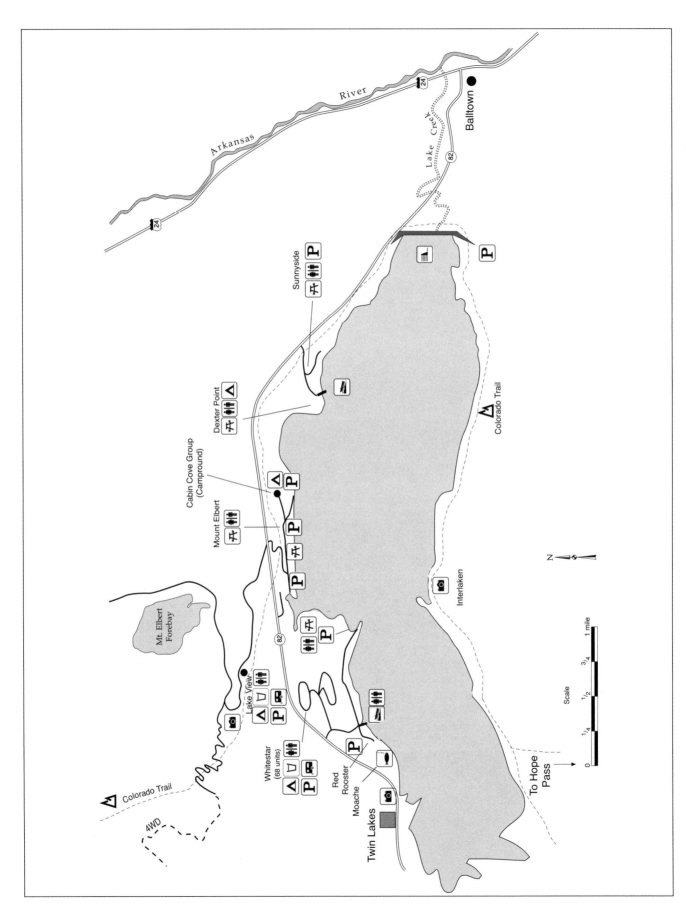

Arkansas River

24

24

Lake Creek

82

Balltown

Sunnyside

Dexter Point

Cabin Cove Group
(Campround)

Mount Elbert

Colorado Trail

Mt. Elbert
Forebay

Interlaken

Lake View

Whitestar
(68 units)

Red
Rooster

Moache

Twin Lakes

82

Colorado Trail

4WD

N

Scale

0   1/4   1/2   3/4   1 mile

To Hope
Pass

DIRECTIONS From Colorado Springs travel west on US 24 to Buena Vista. Follow US 24 north out of Buena Vista for 17 miles to Colorado 82. Drive six miles west on Colorado 82 to the Twin Lakes Reservoirs.

FEE Camping fee. Boat ramp fee.

SIZE 2,440 acres

ELEVATION 9,201 feet.

MAXIMUM DEPTH 70 feet.

FACILITIES Full-service campgrounds.

RECREATION Fishing, boating, hiking. Hiking trailheads include Shore Pretty Overlook (South Mount Elbert and Colorado Trails); Willis Gulch Trailhead (Colorado Trail, Hope Pass and Big Willis Gulch Trails); Black Cloud Trailhead; and Interlocken Trailhead (Interlocken and Colorado Trails).

The majority of the developed recreation areas are located on the north side of Twin Lakes, approximately 20 miles south of Leadville.

BOAT RAMPS Two ramps, both paved.

CAMPING Five campgrounds have 216 total campsites. Fire pits, picnic tables, vault toilets and drinking water are all available. For reservations call (877) 444-6777 or visit www. recreation.gov.

The area around Twin Lakes is closed to camping except in designated areas. Users should note and read other special regulations applicable to the Twin Lakes area, which are posted throughout the recreation area.

There are also seven fishermen parking areas and five picnic areas.

FISH Lake, rainbow and brown trout, and kokanee salmon. Lake trout are caught trolling with sucker meat, Rapala and Flatfish lures. Rainbow trout are caught from shore using salmon eggs and Power Bait. Brown trout are caught trolling with nightcrawlers on Pop Gear or Rapala lures. Kokanee salmon are caught trolling with Red Magics and Kokanee Killers. Mackinaw limit is one fish as part of a four-fish limit. All lake trout between 22 and 34 inches must be returned to water immediately. Check current fishing regulations.

OTHER INFORMATION Twin Lakes are natural glacier-formed lakes which have been enlarged to provide additional storage capacity for the Fryingpan/Arkansas Project of the US Department of Interior, Water and Power Resources Service. Water from the Fryingpan River drainage on the western slope of the Continental Divide is brought through the Boustead Tunnel to Turquoise Reservoir. Then it is transported by conduit to the forebay north shore of Twin Lakes.

During periods of low power use, the turbines are reversed to pump water back into the forebay. Here it is held until the demand for electric power increases. The water is again cycled through the turbines.

Fishing and boating are favorite activities at Twin Lakes. The lakes are especially noted for their mackinaw trout with occasional trophy sized "macks" up to 35 inches in length being taken by anglers. State laws apply to boating and fishing.

The Colorado Trail offers hiking from Twin Lakes, north to Tennessee Pass (30 miles) and south to Cottonwood Creek (25 miles). Buena Vista is east of Cottonwood Creek. The Colorado Trail also passes through the recreation area connecting with auxiliary trails to Willis Lake, Mt. Elbert and Interlaken.

Historic Interlaken, a resort complex in the 1880s, is on the southwest shore of the lower lake. Both Interlaken and Twin Lakes Village have been placed on the National Register of Historic Places.

SPECIAL RESTRICTIONS Keep fires inside stoves, grills, or fireplace rings provided at picnic sites. Do not deface, remove, or destroy plants and trees. Pets must be on a leash not longer than six feet. Drive all motor vehicles only on developed roads within the area. The area is closed to use of firearms or fireworks.

CONTROLLING AGENCY Pike & San Isabel National Forests.

INFORMATION Leadville Ranger District (719) 486-0749.

MAP REFERENCES San Isabel National Forest. Benchmark's *Colorado Road & Recreation Atlas:* p. 86.

Union Reservoir

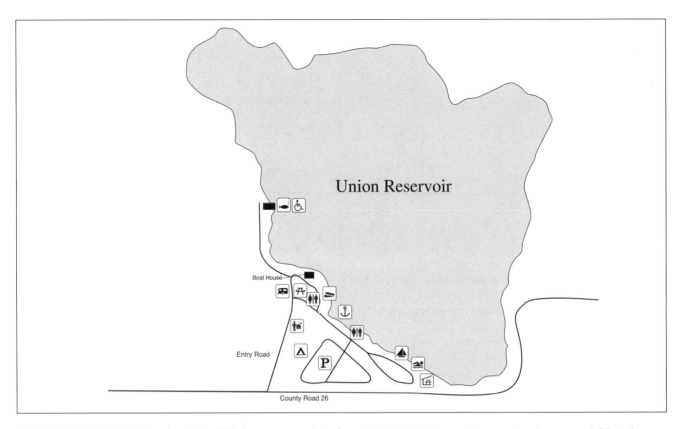

Union Reservoir

Boat House

Entry Road

P

County Road 26

DIRECTIONS From Denver travel north on I-25 to the Longmont exit. Then take Colorado 119 and drive 0.5 miles west to County Road 1. Then head north on County Road 1 for 1.5 miles to the Union Reservoir entrance on the east side of the road.

FEE Daily or annual pass; boat-launching fee.

SIZE 736 acres.

ELEVATION 4,956 feet.

FACILITIES Restrooms, picnic areas, concessions, fishing pier, drinking water, pavilion and fish-cleaning station.

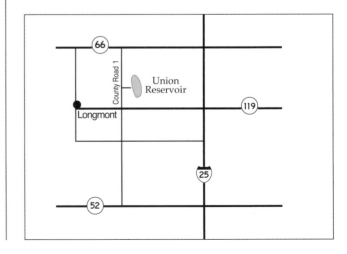

tables and fire rings. Restrooms are nearby. No hookups or dump stations are available. Reservations are recommended.

FISH Crappie, perch, bass, sunfish, tiger muskie, walleye, catfish, wiper and rainbow trout. Crappie: The best area has been along the west shore on jigs and minnows suspended below a bobber. Wiper: The best action has come by trolling with minnows, Rapalas, fishing minnows or worms suspended below a bobber from shore. Rainbow are caught on typical bait. Catfish are caught on dough bait or chicken liver fished on the bottom.

CONTROLLING AGENCY Longmont Parks Division.

BOAT RAMP Two-lane boat ramp with floating dock. Boating is restricted to wakeless speed only.

RECREATION Swim beach with on-duty lifeguard, boating, fishing camping, bird watching, picnicking, sailing and wind surfing.

CAMPING There are 42 primitive campsites with picnic

INFORMATION Union Reservoir Office, (303) 772-1265.

HANDICAPPED ACCESS Union Reservoir is accessible to persons with disabilities. An accessible fishing pier is located on the west shore and two accessible restrooms are available.

MAP REFERENCES Benchmark's *Colorado Road & Recreation Atlas:* p. 63.

Vallecito Reservoir

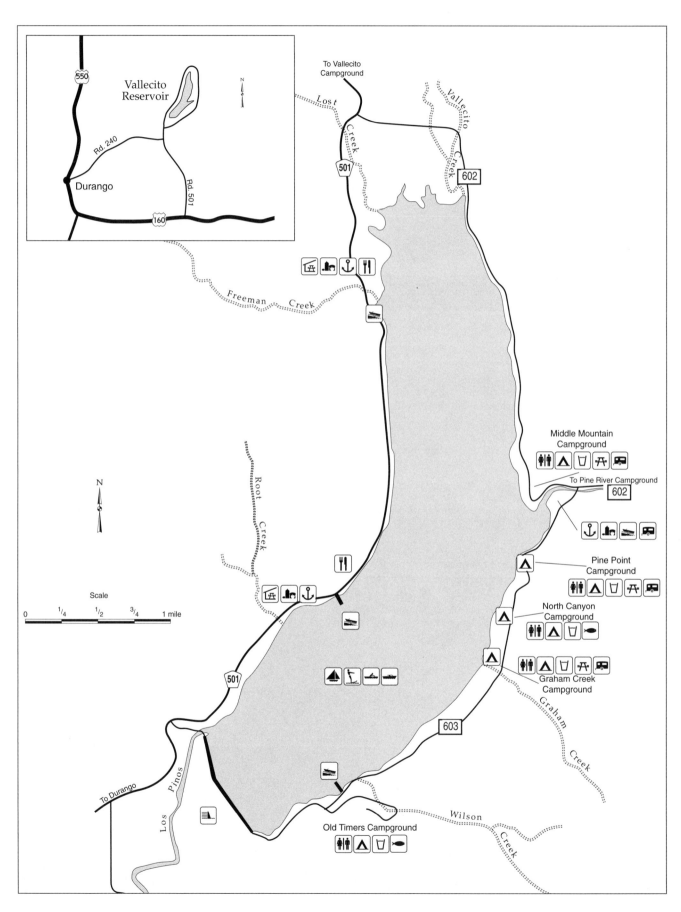

Vallecito Reservoir

550

Rd. 240

Durango

Rd. 501

160

To Vallecito Campground

Lost Creek

Vallecito Creek

501

602

Freeman Creek

N

Root Creek

Scale

0   1/4   1/2   3/4   1 mile

Middle Mountain Campground

To Pine River Campground

602

Pine Point Campground

North Canyon Campground

501

Graham Creek Campground

603

Graham Creek

Los Pinos

To Durango

Old Timers Campground

Wilson Creek

Vallecito Reservoir

DIRECTIONS From Durango travel east on Colorado 160 to Bayfield, then drive north on Colorado 501 for about 12 miles to the reservoir.

FEE Camping and boating fee.

SIZE 2,718 acres.

ELEVATION 7,750 feet.

MAXIMUM DEPTH 125 feet.

FACILITIES Marina, store, full-service campgrounds, boat rentals, gas, groceries, trails and stables.

BOAT RAMP There are two concrete boat ramps.

FISH Vallecito Reservoir is known for record-class brown trout and northern pike. It also has catchable rainbow trout and kokanee salmon. When trolling use rigs, spinners, and fly tackle. Browns are caught primarily by trolling; northerns are taken with large lures or waterdogs. Rainbows and kokanee are caught on Pop Gear and trolling a white or yellow spoon through the summer season.

RECREATION Boating, sailing, hiking, horseback riding, cross-country skiing, snowmobiling, personal watercraft. All water sports.

CAMPING Seven campgrounds offer numerous campsites. They include Graham Creek (25 sites), Middle Mountain (24 sites), North Canyon (21 sites), Pine Point (30 sites), Pine River (6 sites), Vallecito (80 sites) and Old Timers (10 sites). Sites can accommodate units up to 35 feet in length. Vault toilets, picnic tables, fire grates and water are available at all campgrounds. Vallecito Campground, Wapiti Loop, is the only campground accepting reservations. Other sites are on a first-come, first-served basis. Reservations are recommended for holiday weekends. (Pine River and Vallecito Campgrounds not shown on map at left.)

OTHER INFORMATION Vallecito Reservoir, with 22 miles of shoreline, is located in the heart of the great San Juan National Forest in southwestern Colorado. A leisurely 45-minute drive from Durango via Florida Road or a 25-minute drive from Bayfield, the Vallecito Reservoir area offers a multitude of conveniences and excellent accommodations for over 1,000 people.

You can fish in Vallecito Reservoir or stream fish in the Pine and Vallecito Rivers which feed into the lake. Some of the best fishing in the United States is located up and down the rivers. Native, rainbow, and brown trout, northern pike and kokanee salmon are caught in large quantities.

Vallecito Reservoir is the headquarters for backpacking and horseback riding trips into the Weminuche Wilderness and to Emerald Lake. There are public facilities at the Pine River Trailhead and parking for your vehicle.

Deer, elk, bear, bighorn sheep, wild turkey, grouse and other small game are found in abundance at Vallecito. For winter activities, there are miles of cross-country skiing trails to explore.

A few spots of interest include Ignacio, the burial grounds of Ute chiefs, and Gem Village near Bayfield. The Mesa Verde Cliff Dwellings, which date back to AD 1100, are just 37 miles west of Durango on US 160. There is the historic Durango & Silverton Narrow Gauge Railroad which winds through the rugged Las Animas Valley and spectacular mountain scenery between Durango and the famous old mining town of Silverton.

There are also numerous National Forest campgrounds nearby, as well as well-stocked grocery stores, laundromats, service stations, liquor stores, restaurants and cocktail lounges.

Additional accommodations include private campgrounds, cabins and a motel.

CONTROLLING AGENCY Pine River Irrigation District.

INFORMATION Columbine Ranger District (970) 884-2512.

MAP REFERENCES San Juan National Forest map. Benchmark's *Colorado Road & Recreation Atlas:* p. 122.

Vega State Park

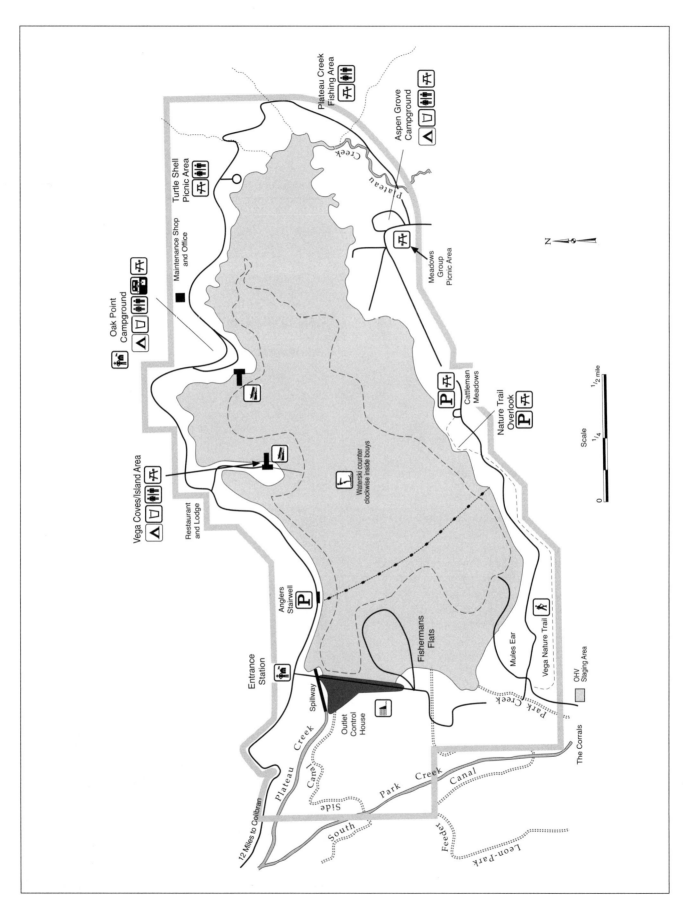

Plateau Creek Fishing Area

Aspen Grove Campground

Turtle Shell Picnic Area

Maintenance Shop and Office

Meadows Group Picnic Area

Oak Point Campground

Cattleman Meadows

Nature Trail Overlook

N

Scale

0   ¼   ½ mile

Vega Coves/Island Area

Restaurant and Lodge

Waterski counter clockwise inside bouys

Anglers Stairwell

Fishermans Flats

Mules Ear

Vega Nature Trail

OHV Staging Area

Entrance Station

Spillway

Outlet Control House

The Corrals

Plateau Creek

Side Canal

South Park Creek Canal

Leon-Park Feeder

Park Creek

12 Miles to Collbran

Vega State Park

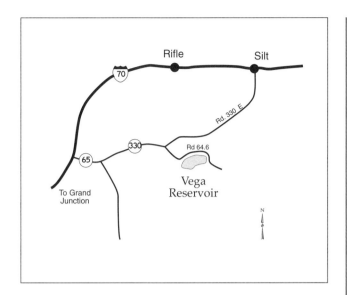

DIRECTIONS Take exit 49 off I-70 to Colorado 65, go about 18 miles on Colorado 65 and turn left onto Colorado 330. Go about 24 miles to the town of Collbran. Stay on 330 East past Collbran for about 8 miles to County Road 64.6 and turn right at the "Vega State Park" sign. Go 5 miles to the Park Entrance Station, and go another half mile and turn right across Vega Dam to the Park Visitor Center.

FEE Daily or annual pass.

SIZE 900 acres.

ELEVATION 7,985 feet.

MAXIMUM DEPTH 94 feet.

FACILITIES Restrooms, picnic tables, visitor center, stables, and fishing piers.

BOAT RAMP Three boat ramps available at Oak Point, Island Area, and Early Settlers Day Use Area.

FISH Few lakes can surpass Vega as a trout fishery. Rainbow trout weighing five pounds have been caught here, and 1.5-pound fish are common. Brook, brown, and cutthroat trout can also be found.

RECREATION Fishing, boating, biking, hunting, OHV riding, waterskiing, wind surfing, personal watercraft. Cold weather enthusiasts come to Vega for ice fishing, ice skating, snowmobiling, cross-country skiing and snow play.

CAMPING There are 109 individual and group campsites in the park that can accommodate both tents and campers. Early Settlers Campground offers 33 sites with electrical and water hookups, flush toilets, and showers. Aspen Grove Campground has 27 sites, and Oak Point Campground has 39 sites. Both offer water pumps and vault toilets. Pioneer Campground offers 10 walk-in sites and 5

Vega State Park

rustic cabins. All sites may be reserved in advance by calling for reservations. A holding tank dump station is located in the Oak Point Campground. There are many individual picnic sites, and the park has one group picnic shelter that accommodates up to 100 people.

SPECIAL RESTRICTIONS Operate vehicles only on designated roads and parking lots. Build campfires only in the fire rings or grills that are provided. Place litter and trash in the receptacles provided. Keep all pets on a leash no more than six feet long.

OTHER INFORMATION Vega Reservoir is in west central Colorado, south of Colorado 330 East. The lake is fed by Plateau Creek and by a diversion off Leon Creek during the spring runoff. The primary use of the lake is for irrigation, so fluctuation in water level occurs. The Spanish word "vega" means meadow. Cattle grazed here from the late 1880s until 1962 when the 900-surface-acre reservoir was completed by the Bureau of Reclamation.

The area surrounding the reservoir became part of Colorado's system of state parks and state recreation areas in 1967 and is administered by the Colorado Division of Parks and Outdoor Recreation. What was once meadow is now a playground for boaters, water skiers, fishermen and other outdoor recreation enthusiasts. Cattle drives still pass through the area in spring and fall, and visitors could be fortunate enough to see one.

Vega is already well known as a winter sports area. Warm weather visitors can expect sub-alpine beauty and mild temperatures characteristic of its 8,000-foot elevation.

Waterskiing is allowed in an area marked by buoys. Waterskiing season usually begins in late June and ends by mid-August depending on water level.

An OHV/snowmobiling area is located in the southwest corner of the park. The off-road vehicle trail leads to Grand Mesa which is especially popular in the fall. In winter it serves as an access route for snowmobilers.

There are also hiking trails. A self-guiding nature trail with interpretive view stations leads walkers through aspen forests and other high-elevation vegetation.

Deer, elk, beaver and many kinds of waterfowl frequent the area. Weasel, rabbits, chipmunks, hawks, blue grouse and the elusive wild turkey are also found here, as well as coyotes and bobcats.

CONTROLLING AGENCY Colorado State Parks.

INFORMATION Collbran Office (970) 487-3407.

HANDICAPPED ACCESS Handicapped accessible restrooms, picnic tables, camping and fishing. Some facilities accessible with assistance.

MAP REFERENCES Benchmark's *Colorado Road & Recreation Atlas:* p. 84.

Williams Fork Reservoir

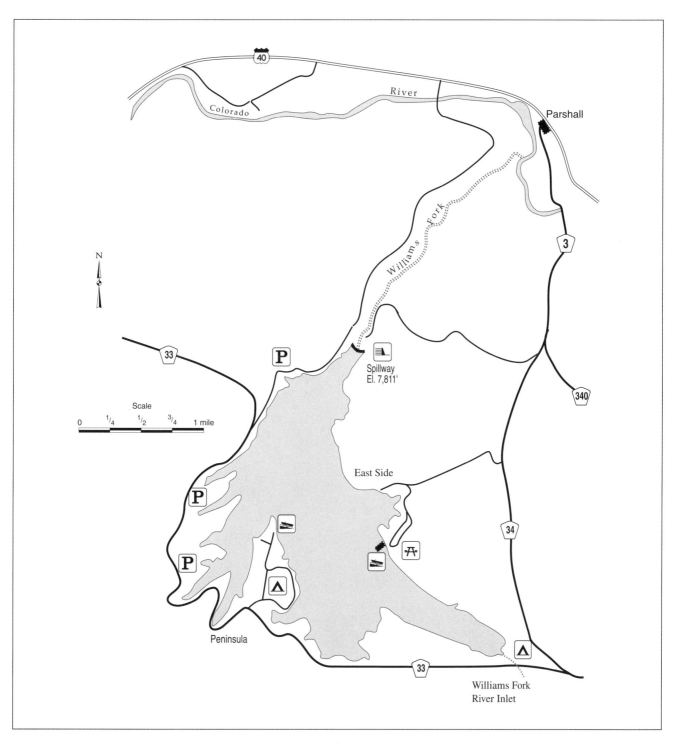

DIRECTIONS From north on Colorado 9 to Kremmling. From Kremmling head east on US 40 to Parshall. Then go south from US 40 about 0.7 miles to the reservoir. There also is access via Forest Road 132 east from Colorado 9, five miles south of Green Mountain Reservoir.

FEE No.

SIZE 1,810 surface acres.

ELEVATION 7,811 feet.

FACILITIES Dump station, picnic tables.

BOAT RAMP Two concrete boat ramps with courtesy docks are available. One is on the east side of the reservoir, the other is located at Peninsula Campground.

FISH Rainbow, lake, and brown trout, kokanee salmon and

Williams Fork Reservoir

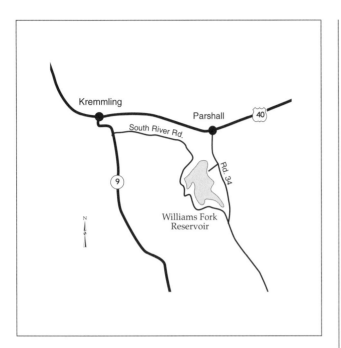

northern pike. For northern pike use jointed Rapalas, spinner baits or large spoons tipped with sucker meat and cast from shore or boats along the shallow areas. Also large

streamer flies can be trailed behind a casting bubble. Trout fishing is good for catchable stocked rainbow and occasional larger fish using baits such as salmon eggs, Power Bait and nightcrawlers along the bottom cast from shore or trolling various types of flashers or gang trolls tipped with a nightcrawler. Troll for kokanee salmon with Cherry Bobbers, red armies lures and Rainbow Needlefish.

RECREATION Camping, boating, fishing and sail boating. Water-contact sports such as swimming are not permitted except wind surfing with full wet suit. No waterskiing, no personal watercrafts are permitted.

CAMPING Camping, which is free, is allowed only in designated areas. Facilities include a dump station on east side, restrooms, fire pits and picnic tables. No water.

CONTROLLING AGENCY Denver Water Department.

INFORMATION Bureau of Land Management, Kremmling Office (970) 724-3000.

MAP REFERENCES Routt National Forest map. Benchmark's *Colorado Road & Recreation Atlas:* p. 61.

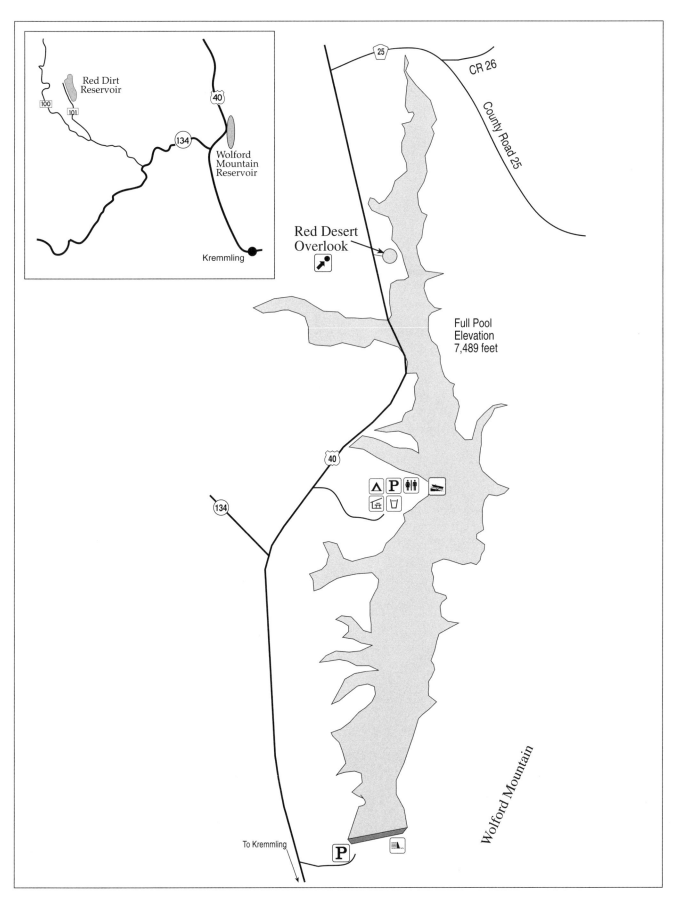

Red Dirt Reservoir

100

101

40

134

Wolford Mountain Reservoir

Kremmling

25

CR 26

County Road 25

Red Desert Overlook

Full Pool Elevation 7,489 feet

40

134

To Kremmling

P

Wolford Mountain

Wolford Mountain Reservoir

DIRECTIONS Take I-70 west to Silverthorne exit. Drive north from Silverthorne on Colorado 9 for 38 miles to Kremmling. Travel west on US 40 from Kremmling five miles to the Ritschard Dam and two miles farther to the Wolford Mountain Recreation Area. There are three access points off US 40 to the reservoir.

FEE Daily or annual pass. Camping fee.

SIZE 1,550 surface acres.

ELEVATION 7,489 feet.

MAXIMUM DEPTH 110 feet.

FACILITIES Picnic tables, shelters, fire grates, boat rental, fish cleaning station, water, electricity, trash removal and a dump station.

BOAT RAMP Concrete boat ramp available at the recreation area.

FISH Rainbow and cutbow trout, kokanee salmon, splake, sucker.

RECREATION Camping, fishing, hiking, biking.

CAMPING One full-service campground has 48 campsites, all with electricity. Restrooms have composting toilets and running water. Two group-use sites are available by reservation only. Day-use area has a shelter, picnic tables and charcoal grills. Campground offers boat rental, waterskiing equipment rental, and picnic sites.

OTHER INFORMATION The reservoir is located on Muddy Creek, about 5 miles northwest of the town of Kremmling. The reservoir extends up Muddy Creek a distance of 5.5 miles and has a surface area of 1,550 acres. The dam site on the west flank of Wolford Mountain is a prominent local feature which consists of Precambrian granite resting upon Cretaceous shales. The shoreline of Wolford Mountain Reservoir consists of Pierre and Niobrara shales, with vegetation consisting of sage brush and dryland grasses.

The Wolford Mountain Reservoir has a capacity of 66,000-acre-feet of water. The Denver Water Department owns 24,000 acre-feet for dry year water supply use. The balance of the water is split between Middle Park Water Conservancy District general western slope demands and environmental and mitigation uses and a conservation pool. The main outlet capacity is 800 cubic feet per second. The service spillway capacity is 24,000 cubic feet per second with an emergency spillway capable of passing an additional 20,000 cubic feet per second.

CONTROLLING AGENCY Colorado River Water Conservation District.

INFORMATION Reservations: (800) 472-4943
Current Conditions: Kremmling Fishin' Hole (970) 724-9407

HANDICAPPED ACCESS Handicapped accessible restrooms and camping.

MAP REFERENCES Benchmark's *Colorado Road & Recreation Atlas:* p. 60.

Yampa River State Park

DIRECTIONS Yampa River State Park offers several public access sites along a 134-mile stretch of the Yampa River between Hayden and the Dinosaur National Monument near the Utah border. The central park office is located on US 40 between the cities of Craig and Hayden. See the map below for locations of all public access sites.

FEE Daily or annual pass.

SIZE 600 acres.

ELEVATION 6,360 feet.

MAXIMUM DEPTH 68 feet.

FACILITIES Restrooms and tables. Eleven miles of recreation trails for non-motorized use. This has area has been a State Park since 1998 and is being developed.

BOAT RAMP A paved boat ramp is located at Elkhead boat launch area. There are 13 access points along the Yampa River, 7 of which have boat ramps.

FISH Smallmouth bass, largemouth bass, mackinaw, northern pike, channel catfish, bluegill, green sunfish, black crappie and sucker. Bag and possession limit 2 bass, bass between 12 and 15 inches must immediately be returned to the water. Mackinaw bag and possession limit is one fish, mackinaw between 22 and 34 inches must immediately be returned to the water. Elkhead Creek above and below the reservoir.

CAMPING There are four options for camping: Headquarters Campground- 50 sites (35 have electric hookup), Yampa River Primitive Camping- 6 sites, undeveloped camping along the Yampa River, and Elkhead Reservoir Campground- 25 sites.

INFORMATION Colorado State Parks. Hayden Office (970) 276-2061.

MAP REFERENCES Benchmark's *Colorado Road & Recreation Atlas:* p. 45.

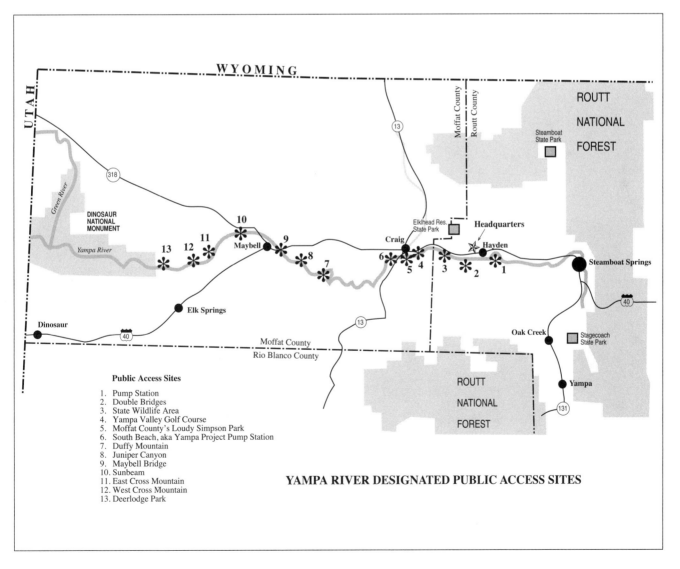

Public Access Sites

1. Pump Station
2. Double Bridges
3. State Wildlife Area
4. Yampa Valley Golf Course
5. Moffat County's Loudy Simpson Park
6. South Beach, aka Yampa Project Pump Station
7. Duffy Mountain
8. Juniper Canyon
9. Maybell Bridge
10. Sunbeam
11. East Cross Mountain
12. West Cross Mountain
13. Deerlodge Park

YAMPA RIVER DESIGNATED PUBLIC ACCESS SITES

Yampa River State Park

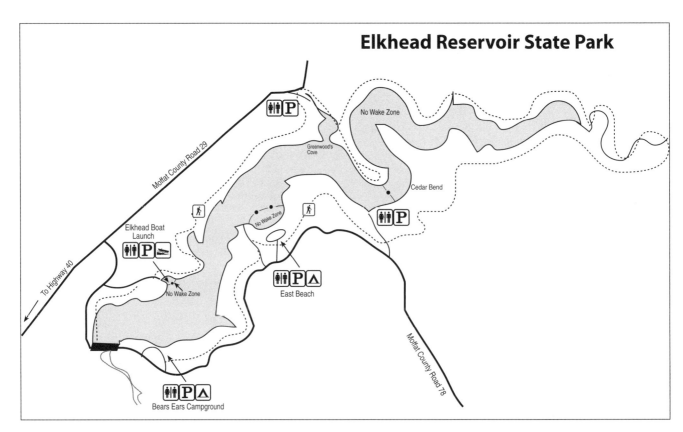

Elkhead Reservoir State Park

No Wake Zone

Greenwood's Cove

Cedar Bend

Moffat County Road 29

Elkhead Boat Launch

No Wake Zone

To Highway 40

No Wake Zone

East Beach

Moffat County Road 78

Bears Ears Campground

Section 2

Metro and Mountain Lakes and Reservoirs

Index to Metro and Mountain Lakes

Metro/Mountain Lakes are smaller lakes that you can reach in one day from Metro Denver. Many lakes are only accessible by 4-wheel drive vehicles or on foot. Most lakes have boat restrictions. The restrictions are in the text. Elevations of the lakes are from 7,200 feet to over 12,000 feet, so prepare for rapid weather changes. The mountain lakes are divided into five maps. Each lake is numbered and is accompanied by descriptive text. The descriptions of the mountain lakes include general location, size and maximum depth, types of fish present, name of managing agency, and comments including USGS topo map, elevation, and any other information specific to the lake.

Metro and Mountain Lakes

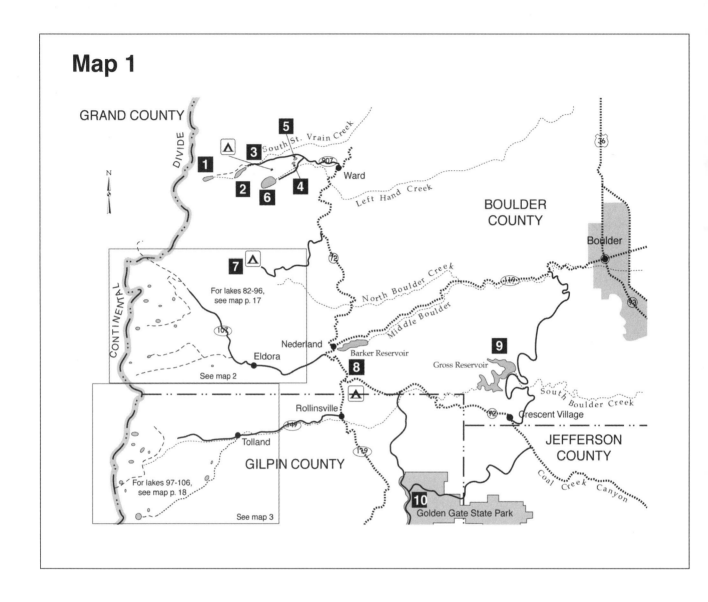

Map 1

GRAND COUNTY

DIVIDE

CONTINENTAL

N

1 **2** **3** **4** **5** **6**

South St. Vrain Creek

Ward

Left Hand Creek

BOULDER COUNTY

Boulder

7

For lakes 82-96, see map p. 17

North Boulder Creek

Middle Boulder

Nederland

Barker Reservoir

8

9

Gross Reservoir

South Boulder Creek

Eldora

See map 2

Rollinsville

Crescent Village

JEFFERSON COUNTY

Tolland

GILPIN COUNTY

For lakes 97-106, see map p. 18

See map 3

Coal Creek Canyon

10

Golden Gate State Park

MAP 1

1 Lake Isabelle
Location: Boulder County. Arapaho/Roosevelt National Forest. From Ward, go north one-quarter mile on Colorado 72 to Brainard Lake Road. Go west about 4 miles to Brainard Lake. Drive past Brainard Lake to trailhead. Hike west about one-quarter mile on Pawnee Pass Trail to Long Lake. Continue west on the trail (along north shore of Long Lake) about 2 miles to Lake Isabelle.
Size: 30 acres; 40 feet maximum depth.
Fish: Rainbow trout.
Agency: USFS-Boulder Ranger District.
Comments: USGS Ward quad; elevation 10,868 feet. In Indian Peaks Wilderness Area. Camping at Pawnee campground near Brainard Lake. Non-motorized boats only.
Special Regulations: 1. Fishing by artificial flies or artificial lures only; 2. Bag, possession and size limit for trout is 2 fish.

2 Long Lake
Location: Boulder County. Arapaho/Roosevelt National Forest. From Ward, go north one-quarter mile on Colorado 72 to Brainard Lake Road. Go west about 4 miles to Brainard Lake. Drive past Brainard Lake to trailhead. Hike west about one-quarter mile on Pawnee Pass Trail to Long Lake.
Size: 39.5 acres; 22 feet maximum depth.
Fish: Rainbow, cutthroat, and brook trout.
Agency: USFS-Boulder Ranger District.
Comments: USGS Ward quad; elevation 10,500 feet. Camping at Pawnee campground. Non-motorized boats only.
Special Regulations: 1. Fishing by artificial flies or artificial lures only; 2. Bag, possession and size limit for trout is 2 fish. 3. Fishing is prohibited in the outlet stream of Long Lake to the bridge at the inlet at Brainard Lake from May 1 through July 15, as posted.

3 Brainard Lake
Location: Boulder County. Arapaho/Roosevelt National Forest. From Ward, go north one-quarter mile on Colorado 72 to Brainard Lake Road. Go west about 4 miles to Brainard Lake.
Size: 14 acres; 8 feet maximum depth.
Fish: Rainbow trout (catchable size stocked), brook and brown trout.
Agency: USFS-Boulder Ranger District.
Comments: USGS Ward quad; elevation 10,350 feet. Camping at Pawnee campground. Non-motorized boats only. Fee Area except from early Oct.-mid-June.

4 Moraine Lake
Location: Boulder County. Arapaho/Roosevelt National Forest. From Ward, go north on Colorado 72 one-quarter mile to Brainard Lake Road. Go west 2 miles past the turnoff for Lefthand Park. Moraine Lake is on the south beyond Red Rock Lake.
Size: 2.0 acres; 3.5 feet maximum depth.
Fish: Rainbow trout.
Agency: USFS-Boulder Ranger District.
Comments: USGS Ward quad; elevation 10,150 feet. Non-motorized boats only.

5 Red Rock Lake
Location: Boulder County. Arapaho/Roosevelt National Forest. From Ward, go north on Colorado 72 one-quarter mile to Brainard Lake Road. Go west 2 miles past the turnoff for Lefthand Park. On the south side of the road is the parking area for Red Rock Lake.
Size: 5.1 acres; 3 feet maximum depth.
Fish: Rainbow (catchable size stocked), cutthroat and brook trout.
Agency: USFS-Boulder Ranger District.
Comments: USGS Ward quad; elevation 10,300 feet. Grassy
with lily pads. Non-motorized boats only.

6 Lefthand Creek Reservoir
Location: Boulder County. Arapaho/Roosevelt National Forest. From Ward, go north on Colorado 72 one-quarter mile to Brainard Lake Road. Go west 2 miles to Lefthand Park. Follow road southwest about 2 miles to the reservoir.
Size: 100 acres; 34 feet maximum depth.
Fish: Rainbow, brook, brown trout and splake.
Agency: USFS-Boulder Ranger District.
Comments: USGS Ward quad; elevation 10,600 feet. Boats allowed, no gasoline motors.

7 Rainbow Lakes
Location: Boulder County. Arapaho/Roosevelt National Forest. From Ward, go south about 3.5 miles on Colorado 72. Turn west on gravel road. (First fork goes to the University of Colorado Camp.) Take south fork about 4 miles to Rainbow Lakes Campground and Trailhead.
Size: Ten beaver ponds; from 1 to 4 acres; 15 feet maximum depth.
Fish: Brook, cutthroat, and rainbow trout.
Agency: USFS-Boulder Ranger District.
Comments: USGS-Boulder quad; elevation 10,200 feet. Forest Service campground. Trailhead for Arapaho Glacier Trail. Non-motorized boats only. Fishing period July-Sept.

8 Barker Reservoir
Location: Boulder County. Arapaho/Roosevelt National Forest. From Nederland, go one-half mile east along Colorado 119. The reservoir is on Middle Boulder Creek. From Boulder, go 15 miles west on Colorado 119.
Size: 380 acres; 100 feet maximum depth.
Fish: Rainbow, (catchable sized stocked), brook and brown trout, sucker and splake.
Agency: Public Service Company.
Comments: USGS Nederland, Tungsten quads. Fluctuation water storage reservoir. No boats. No ice fishing.

Metro and Mountain Lakes

9 Gross Reservoir

Location: Boulder County. Arapaho/Roosevelt National Forest. From Boulder, drive west on Baseline Road to Flagstaff Mountain Road. Wind southwest for about 7 miles to Gross Reservoir. From Golden, drive 18 miles north on Colorado 93 to Boulder and turn west on Baseline Road. Follow directions from Boulder. Alternate route from Golden, drive 7 miles north on Colorado 93, then turn west on Colorado 72, Coal Creek Canyon. Drive about 10 miles northwest to Crescent Village. Drive North about 3 miles to Gross Reservoir. From Nederland, drive about 2.5 miles south on Colorado 119. Turn east on Colorado 72, and drive about 9 miles east to Crescent Village. Turn north and go about 3 miles to Gross Reservoir.

Size: 440 acres; 230 feet maximum depth.

Fish: Rainbow trout (catchable size stocked), brook, brown and lake trout, kokanee salmon, splake, and tiger muskie.

Agency: Denver Water Board.

Comments: USGS Tungsten and Eldorado Springs quads; elevation 7,287 feet. Steep banked. Camping on USFS land only; no boats or floating devices. Fires in firepits on north side only. Open 4 AM to 9 PM. Ice fishing at your own risk. Water level fluctuation.

Special Regulations: Kokanee salmon snagging permitted September 1 to January 31. West side of reservoir is closed from around December 1 to May 15 for critical wildlife habitat.

10 Golden Gate State Park Lakes

Location: Gilpin and Jefferson Counties. From Golden, drive north one mile on Colorado 93 to Golden Gate Canyon Road, then west 14 miles to the park. From Nederland, drive about 8 miles south on Colorado 119 to park entrance. Alternate route from Boulder, drive south on Colorado 93 to Golden Gate Canyon Road, and turn west 14 miles to the park. Also reached from Colorado 119 by turning east on Colorado 46 about 5 miles north of Black Hawk.

Size: 9 ponds; 15 acres total,

Fish: Rainbow trout (catchable size stocked), brown and brook trout.

Agency: Colorado Division of Parks & Outdoor Recreation.

Comments: USGS Blackhawk and Ralston Buttes quads; elevation ranges from 7,400 fcct to 10,400 fcct No fishing in visitor center show pond. Park is a fee area. No boats. Ice fishing allowed on Kriley and Slough Ponds. There are 156 campsites available. Bag limit is 4 fish.

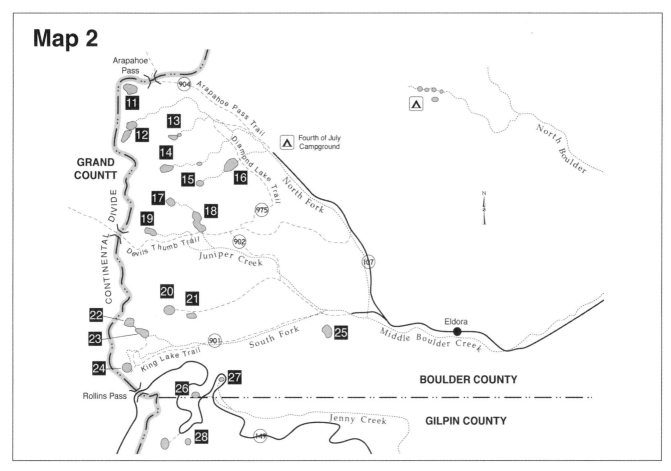

Map 2

11 Lake Dorothy

Location: Boulder County. Arapaho/Roosevelt National Forest. From Nederland, go one-half mile south on Colorado 72/119. Turn west, go 4 miles to Eldora, continue 1 mile and take the north fork of the road, along the north fork of Middle Boulder Creek. Drive about 4 miles to the Fourth of July Campground. Hike north and northwest on Arapaho Pass Trail for about 3 miles. Lake Dorothy is in open country about one-half mile southwest of the summit.
Size: 16 acres; 100 feet maximum depth.
Fish: Cutthroat trout.
Agency: USFS-Boulder Ranger District.
Comments: USGS Monarch Lake quad; elevation 12,100 feet. In Indian Peaks Wilderness Area. Non-motorized boats only.

12 Neva Lakes

Location: Boulder County. Arapaho/Roosevelt National Forest. From Nederland, go one-half mile south on Colorado 72-119. Turn west, go 4 miles to Eldora, continue one mile and take the north fork of the road, along the north fork of Middle Boulder Creek. Drive about 4 miles to the Fourth of July Campground. Hike north on Arapaho Pass Trail. After one-quarter mile take the west fork, Diamond Lake Trail. Go northwest about one-half mile where the trail crosses a small stream and then curves southwest. Cut off at the second stream, north fork of Mid-

dle Boulder Creek, and follow it cross-country for 2 miles. There is no established trail. Use of topo map is advised.
Size: Upper Lake (8.6 acres, 61 feet maximum depth), Lower Lake (10.0 acres, 49 feet maximum depth)
Fish: Cutthroat trout.
Agency: USFS Boulder Ranger District.
Comments: USGS Monarch Lake quad; elevation 11,800 feet, non-motorized boats only.

13 Diamond Lake—Upper

Location: Boulder County. Arapaho/Roosevelt National Forest. From Nederland, go one-half mile south on Colorado 72/119. Turn west, go 4 miles to Eldora, continue one mile and take the north fork of the road, along the north of Middle Boulder Creek. Drive about 4 miles to the Fourth of July Campground. Hike west one-half mile to the Diamond Lake Trail. Hike west one-half mile to the Diamond Lake Trail. Follow Trail 904 and 975 to Diamond Lake. The trail is obscure upstream to upper Diamond.
Size: 6 acres; 17 feet maximum depth.
Fish: Cutthroat anad rainbow trout.
Agency: USFS-Boulder Ranger District.
Comments: USGS East Portal quad; elevation 11,720 feet. Non-motorized boats only.

14 Deep Lake

Location: Boulder County. Arapaho/Roosevelt Na-

Metro and Mountain Lakes

tional Forest. From Nederland, go one-half mile south on Colorado 72/119. Turn west, go 4 miles to Eldora, continue 1 mile and take the north fork of the road, along the north fork of Middle Boulder Creek. Drive about 4 miles to the Fourth of July Campground. Hike north on Arapaho Pass Trail about one-quarter mile and take the west fork, Diamond Lake Trail. Go northwest about one-half mile. Trail crosses a small stream and then curves southwest. Cut off at the second stream, north fork of Middle Boulder Creek, and follow it for three-quarters of a mile. Head west at tributary, follow for three-quarters mile.
Size: 4.9 acres; 20 feet maximum depth.
Fish: Cutthroat and rainbow trout.
Agency: USFS-Boulder Ranger District.
Comments: USGS East Portal and Monarch Lake quads; elevation 11,320 feet. Rough country. Non-motorized boats only.

15 Banana Lake
Location: Boulder County. Arapaho/Roosevelt National Forest. From Nederland, go one-half mile south on Colorado 72/119. Turn west, go 4 miles to Eldora, continue one mile and take the north fork of the road, along the north fork of Middle Boulder Creek. Drive about 4 miles to the Fourth of July Campground. Follow Trail 904 and 975 to Diamond Lake, then follow obscure trail upstream for one-half mile. There is no established trail. Use of topo map is advised.
Size: 1.5 acres; 13 feet maximum depth.
Fish: Cutthroat trout.
Agency: USFS-Boulder Ranger District.
Comments: USGS East Portal quad; elevation 11,320 feet. No boats.

16 Diamond Lake
Location: Boulder County. Arapaho/Roosevelt National Forest. From Nederland, go one-half mile south on Colorado 72/119. Turn west, go 4 miles to Eldora, continue 1 mile and take the north fork of the road, along the north fork of Middle Boulder Creek. Drive about 4 miles to the Fourth of July Campground. Follow Forest Service Trail 904 and 975 to the lake.
Size: 14 acres; 17 feet maximum depth.
Fish: Brook, rainbow, cutthroat and lake trout.
Agency: USFS-Boulder Ranger District.
Comments: USGS East Portal quad; elevation 10,920 feet. In Indian Peaks Wilderness Area. Non-motorized boats only.

17 Storm Lake
Location: Boulder County. Arapaho/Roosevelt National Forest. From Nederland, go one-half mile south on Colorado 72/119. Turn west, and go 4 miles to Eldora, continue 1 mile on Forest Road 109 to the Hessie Trailhead. Walk west about 5 miles on Trail 902 to Jasper Lake. From north the end of Jasper, follow inlet stream a half mile northwest to Storm Lake.
Size: 7 acres; 22 feet maximum depth.
Fish: Native trout.

Agency: USFS-Boulder Ranger District.
Comments: USGS East Portal quad; elevation 11,440 feet. Non-motorized boats only.

18 Jasper Lake
Location: Boulder County. Arapaho/Roosevelt National Forest. From Nederland, go one-half mile south on Colorado 72/119. Turn west, and go 4 miles to Eldora. Continue one mile on Forest Road 109 to the Hessie Trailhead. Walk west about 1.5 miles, then take the north fork. Go up the trail one mile past mine diggings and take the north fork again. Continue north and west for 2.5 miles to Jasper Lake.
Size: 19 acres; 36 feet maximum depth.
Fish: Brook, brown and cutthroat trout.
Agency: USFS-Boulder Ranger District.
Comments: USGS Nederland and East Portal quads; elevation 10,814 feet. In Indian Peaks Wilderness Area. Non-motorized boats only. Water level fluctuates.

19 Devil's Thumb Lake
Location: Boulder County. Arapaho/Roosevelt National Forest. From Nederland go half a mile south on Colorado 72/119. Turn west and go 4 miles to Eldora. Continue one mile on Forest Road 109 to the Hessie Trailhead. Walk west about 1.5 miles then take the north fork. Go up the trail one mile past mine diggings and take the north fork again. Continue north and west for 2.5 miles to Jasper Lake. Follow the trail west for one mile to Devil's Thumb Lake.
Size: 9.8 acres; 37 feet maximum depth.
Fish: Cutthroat, brook, rainbow, and brown trout.
Agency: USFS-Boulder Ranger District.
Comments: USGS East Portal quad; elevation 11,260 feet. In Indian Peaks Wilderness Area. Non-motorized boats only.

20 Skyscraper Reservoir
Location: Boulder County. Arapaho/Roosevelt National Forest. From Nederland, go half a mile south on Colorado 72/119. Turn west, and go 4 miles to Eldora. Continue one mile on Forest Road 109 to the Hessie Trailhead. Hike west about 1.5 miles then take the north fork. Go up the trail one mile past mine diggings to the next fork. Go west 2 miles to Woodland Lake. Go a few hundred yards farther to Skyscraper Reservoir.
Size: 12.9 acres; 28 feet maximum depth.
Fish: Cutthroat trout.
Agency: USFS-Boulder Ranger District.
Comments: USGS East Portal quad; elevation 11,221 feet. In Indian Peaks Wilderness Area. Non-motorized boats only.

21 Woodland Lake
Location: Boulder County. Arapaho/Roosevelt National Forest. From Nederland, go half a mile south on Colorado 72/119. Turn west, and go 4 miles to Eldora. Continue one mile on Forest Road 109 to the Hessie Trailhead. Hike west about 1.5 miles then take the north fork.

Go up the trail one mile past mine diggings to the next fork. Go west 2 miles to Woodland Lake.
Size: 6.5 acres; 7 feet maximum depth.
Fish: Cutthroat trout.
Agency: USFS-Boulder Ranger District.
Comments: USGS East Portal quad; elevation 10,972 feet. In Indian Peaks Wilderness Area. Non-motorized boats only. Camping allowed with a Woodland Travel Zone permit.

22 Bob Lake
Location: Boulder County. Arapaho/Roosevelt National Forest. From Nederland, go half a mile south on Colorado 72/119. Turn west, and go 4 miles to Eldora. Continue one mile on Forest Road 109 to the Hessie Trailhead. Walk 5 miles west on Trail 901, then turn southwest at the fork and follow the south fork of Middle Boulder Creek. At intersection with the north/south trail go north for about a half mile to Betty Lake. From the trail along the west shore of Betty Lake continue an eighth of a mile to Bob Lake. The lake is also accessible by hiking north from Rollins Pass.
Size: 6.5 acres; 71 feet maximum depth.
Fish: Cutthroat trout.
Agency: USFS-Boulder Ranger District.
Comments: USGS East Portal quad; elevation 11,600 feet. In Indian Peaks Wilderness Area. Non-motorized boats only. Only the eastern shore is accessible for fishing.

23 Betty Lake
Location: Boulder County. Arapaho/Roosevelt National Forest. From Nederland, go a half mile south on Colorado 72/119. Turn west, and go 4 miles to Eldora. Continue one mile on Forest Road 109 to the Hessie Trailhead. Walk 5 miles west, then head southwest at the fork and follow the south fork of Middle Boulder Creek. At the intersection with the north/south trail go north about a half mile to Betty Lake. The lake is also accessible by hiking north from Rollins Pass at Needle Eye Tunnel.
Size: 5.8 acres; 11 feet maximum depth.
Fish: Cutthroat trout.
Agency: USFS-Boulder Ranger District.
Comments: USGS Nederland and East Portal quads; elevation 11,500 feet. In Indian Peaks Wilderness Area. Non-motorized boats only.

24 King Lake
Location: Boulder County. Arapaho/Roosevelt National Forest. From Nederland, go a half mile south on Colorado 72/119. Turn west, and go 4 miles to Eldora. Continue one mile on Forest Road 109 to the Hessie Trailhead. Walk 5 miles west then go southwest at the fork and follow the south fork of Middle Boulder Creek. At intersection with the north/south trail go south about a quarter mile to King Lake. The lake is also accessible by hiking north a half mile from Rollins Pass Summit.
Size: 10.4 acres; 65 feet maximum depth.
Fish: Cutthroat and rainbow trout.
Agency: USFS-Clear Creek Ranger District.

Comments: USGS Nederland and East Portal quads; elevation 11,431 feet. In Indian Peaks Wilderness Area. Non-motorized boats only.

25 Lost Lake
Location: Boulder County. Arapaho/Roosevelt National Forest. From Nederland, go a half mile south on Colorado 72/119. Turn west, and go 4 miles to Eldora. Continue one mile on Forest Road 109 to the Hessie Trailhead. Hike west on Trail 902, then head southwest at the fork. Total distance is less than 2 miles.
Size: 4.9 acres; 14 feet maximum depth.
Fish: Rainbow and brook trout.
Agency: USFS-Boulder Ranger District.
Comments: USGS Nederland quad; elevation 9,740 feet. Non-motorized boats only.

26 Jenny Lake
Location: Boulder County. Arapaho/Roosevelt National Forest. From Rollinsville, drive west about 7 miles on Forest Road 149. Before East Portal, turn north on old railroad grade and drive up Rollins Pass Road (Forest Road 149) to Yankee Doodle Lake. Drive one mile farther to Jenny Lake.
Size: 4.5 acres; 9 feet maximum depth.
Fish: Brook and rainbow trout (catchable size stocked), sucker.
Agency: USFS-Boulder Ranger District.
Comments: USGS East Portal quad; elevation 10,917 feet. Forest Road 149 (Rollins Pass Road) is closed at Needle Eye Tunnel.

27 Yankee Doodle Lake
Location: Boulder County. Arapaho/ Roosevelt National Forest. From Rollinsville, drive west about 7 miles on Forest Road 149. Before East Portal, turn north on old railroad grade and drive about 10 miles up Rollins Pass Road (Forest Road 149) to Yankee Doodle Lake.
Size: 5.7 acres; 24 feet maximum depth.
Fish: Rainbow trout (catchable size stocked), and brook trout.
Agency: USFS-Boulder Ranger District.
Comments: USGS East Portal quad; elevation 10,711 feet. Inquire locally about current road conditions. Boats allowed. Forest Road 149 (Rollins Pass Road) is closed at Needle Eye Tunnel.

28 Forest Lakes
Location: Boulder County. Arapaho/Roosevelt National Forest. From Rollinsville, drive west about 8 miles on Forest Road 149 to East Portal. Park by tunnel. On the south side of the road walk through gate, which may be locked. Hike one mile southwest to the intersection with South Boulder Creek Pack Trail. Turn north and go up the trail 1.5 miles. At the half-mile point the trail crosses the creek. Where the creeks meet, take the north fork. Hike a half mile north cross country to the first lake. Continue another half mile to the second. The lake is also accessible by hiking south a half mile from Rollins Pass Road (FR 149)

south of Jenny Lake.
Size: Upper (4.3 acres, 8.2 feet maximum depth); Lower (2.7 acres, 3.2 feet maximum depth).
Fish: Brook trout.
Agency: USFS-Boulder Ranger District
Comments: USGS East Portal quad; elevation 10,800 feet. Small forest ponds. Non-motorized boats only.

29 Arapaho Lakes
Location: Gilpin County. Arapaho/Roosevelt National Forest. From Rollinsville, drive west about 8 miles on Forest Road 149 to East Portal. Park by tunnel. On the south side of the road walk through gate, which may be locked. Hike one mile southwest to intersection with South Boulder Creek Trail. Turn north and go 1.5 miles. At the half-mile point the trail crosses creek. Where the creeks meet, take the west fork. Follow the stream three-quarters of a mile west to the first lake. There are 2 more lakes to the west directly upstream. (After taking the west fork, at about one-eighth mile a small tributary comes in from the south, there is a small pond about one-half mile upstream.)
Size: Lake #1—East (9.8 acres; 49.2 feet maximum depth); Lake #1—Middle (4 acres; 15 feet maximum depth); Lake #2— West (2.5 acres; 16.4 feet maximum depth; no fish).
Fish: Cutthroat trout.
Agency: USFS-Boulder Ranger District.
Comments: USGS East Portal quad; elevation 11,580 feet. Non-motorized boats only.

30 Crater Lakes
Location: Gilpin County. Arapaho/Roosevelt National Forest. From Rollinsville, drive west about 8 miles on Forest Road 149 to East Portal. Park by tunnel. On the south side of the road walk through gate, which may be locked. Hike one mile southwest to the intersection with South Boulder Creek Trail. Turn south into the clearing. Hike west, upstream for a half mile to the first lake. There is no established trail. Uuse of a topo map is advised.
Size: East (5.7 acres; 4 feet maximum depth); Southeast (8.6 acres; 29 feet maximum depth); Middle (14.0 acres; 14 feet maximum depth); West (8.5 acres; 58 feet maximum depth).
Fish: Rainbow, brook, brown and cutthroat trout.
Agency: USFS-Boulder Ranger District.
Comments: USGS East Portal quad; elevation 11,000 feet. Non-motorized boats only.
Helpful Hints: The most productive way to fish these four lakes is from a float tube. A tremendous fly ant hatch (size 14–16) occurs in August on the lower three lakes. The best time to fish the upper lake for cutthroats is when the ice melts in late June or early July. At this time cutthroats often cruise the shallow shoals looking for food and spawning habitat.

31 Clayton Lake
Location: Gilpin County. Arapaho/Roosevelt National Forest. From Rollinsville, drive west about 8 miles on Forest Road 149 to East Portal. Park by tunnel. On the south side of the road walk through gate, which may be locked. Hike one mile southwest to the intersection with South Boulder Creek Trail. Turn south, cross creek, and

hike about one mile to another creek. Turn west and follow the creek cross country for a half mile to Clayton Lake. Use of a topo map is advised.
Size: 5 acres; 4 feet maximum depth.
Fish: Cutthroat trout.
Agency: USFS-Boulder Ranger District.
Comments: USGS East Portal quad; elevation 11,560 feet. Non-motorized boats only.

32 Iceberg Lakes
Location: Gilpin County. Arapaho/Roosevelt National Forest. From Rollinsville, drive west about 8 miles on Forest Road 149 to East Portal. Park by tunnel. On the south side of the road walk through gate, which may be locked. Hike one mile southwest to the intersection with South Boulder Creek Trail. Turn south, cross creek, and hike about one mile to another creek. Turn west and follow the creek cross country for a half mile to Clayton Lake. Continue a quarter mile west along the creek, then take either fork another quarter mile to the lakes.
Size: North Lake (10 acres; 77 feet maximum depth); South Lake (6 acres; 100 feet maximum depth).
Fish: Cutthroat trout.
Agency: USFS-Boulder Ranger District.
Comments: USGS East Portal quad; elevation 11,500 feet. Non-motorized boats only. No fish in south lake.

33 Heart Lake
Location: Gilpin County. Arapaho/Roosevelt National Forest. From Rollinsville, drive west about 8 miles on Forest Road 149 to East Portal. Park by tunnel. On the south side of the road walk through gate, which may be locked. Hike one mile southwest to intersection with South Boulder Creek Trail. Turn south, cross creek, and hike about one mile. Cross another creek and continue south three-quarters of a mile; follow the trail west for a half mile. After the trail crosses a stream, hike west following the stream for a quarter mile to Heart Lake.
Size: 17 acres; 52 feet maximum depth.
Fish: Cutthroat trout.
Agency: USFS-Boulder Ranger District.
Comments: USGS East Portal quad; elevation 12,218 feet. Trail continues west over Roger Pass. Non-motorized boats only.

34 Rogers Pass Lake
Location: Gilpin County. Arapaho/Roosevelt National Forest. From Rollinsville, drive west about 8 miles on Forest Road 149 to East Portal. Park by tunnel. On the south side of the road walk through gate, which may be locked. Hike one mile southwest to intersection with South Boulder Creek Trail. Turn south, cross creek, and hike about one mile. Cross another creek and continue south three-quarters of a mile; follow the trail west for a half mile. Trail crosses stream. Stay on this trail for a short distance until you can see Rogers Pass Lake just to the south of the trail. It is before Rogers Pass.
Size: 5.6 acres; 6.5 feet maximum depth.
Fish: Cutthroat trout.
Agency: USFS-Boulder Ranger District.
Comments: USGS Empire quad; elevation 11,200 feet. Non-motorized boats only.

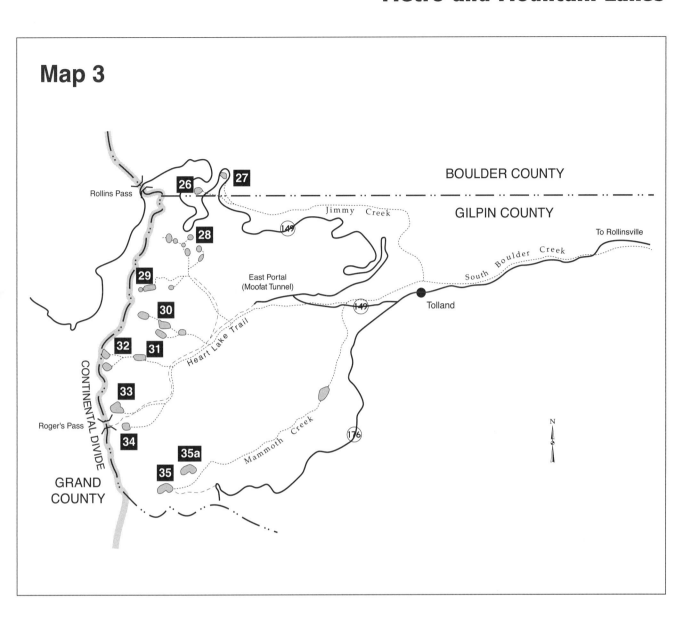

Map 3

BOULDER COUNTY

GILPIN COUNTY

Rollins Pass

26

27

Jimmy Creek

149

To Rollinsville

28

South Boulder Creek

29

East Portal
(Moofat Tunnel)

30

149

Tolland

Heart Lake Trail

32   31

33

CONTINENTAL DIVIDE

Mammoth Creek

176

N

Roger's Pass

34

35a

35

GRAND
COUNTY

---

35 **James Peak Lake**
Location: Gilpin County. Arapaho/Roosevelt National Forest. From Rollinsville, drive west 5.5 miles on Forest Road 149 to Tolland. Turn left on Forest Road 176. Go about 1.5 miles to triple fork. Take right fork toward Kingston Peak and James Peak. It is about 4 miles to the James Peak Trailhead. Hike down the trail for about one mile west to James Peak Lake.
Size: 10 acres; 10 feet maximum depth.
Fish: Cutthroat trout.
Agency: USFS-Boulder Ranger District.
Comments: USGS Empire and East Portal quads; elevation 11,100 feet. Non-motorized boats only.

35a **Little Echo Lake**
Location: Gilpin County, Arapaho/Roosevelt National Forest. From Rollinsville, drive west 5.5 miles on Forest Road 149 to Tolland. Turn south on Forest Road 176. Go about 1.5 miles to triple fork. Take west fork towards Kingston Peak and James Peak., It is about 4.0 miles to James Peak Trailhead. Take Trail #804 about 1.5 miles west to Little Echo Lake. The lake is also accessible from Apex.
Size: 13 acres; 96 feet maximum depth.
Fish: Rainbow and lake trout.
Agency: Central City
Comments: USGS Empire and East Portal quads; elevation 11,100 feet. Non motorized boats only.
Helpful Hints: This lake has a naturally sustained lake trout population. Although most of these fish are smaller (up to 14 inches). A unique angling opportunity exists to catch them on dry flies. Matching a spectacular flying ant hatch (size 14-16) in August while fishing from a float tube can be effective. Fish often feed in the middle of the lake.

Metro and Mountain Lakes

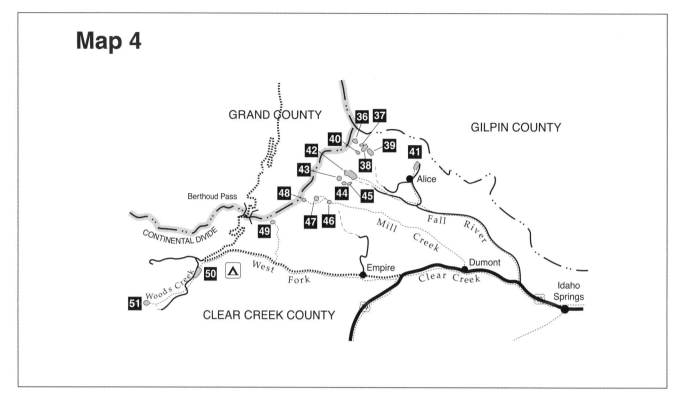

Map 4

GRAND COUNTY

GILPIN COUNTY

36 37

40

42

43

39

41

38

Alice

48

44 45

Berthoud Pass

CONTINENTAL DIVIDE

49

47 46

Mill Creek

Fall River

50

West Fork

Empire

Dumont

Clear Creek

51

Woods Creek

Idaho Springs

CLEAR CREEK COUNTY

---

36 **Ice Lake**
Location: Clear Creek County. Arapaho/Roosevelt National Forest. Travel west on I-70 about 2 miles past Idaho Springs and take exit 238, Fall River Road. Drive north on Fall River Road for 9 miles to the town of Alice. Turn left on Silver Creek then turn right on Texas Drive past the Glory Hole Mine. Take the first dirt road right after the mine. It is a rough road. Past Steuart Lake, hike 0.2 miles to Ohman Lake. Ice Lake is s.2 mile past Ohman Lake.
Size: l2 acres; 102 feet maximum depth.
Fish: Native trout.
Agency: USFS-Clear Creek Ranger District.
Comments: USGS Empire quad; elevation 12,200 feet. Ice stays very late. Non-motorized boats only.

37 **Steuart Lake**
Location: Clear Creek County. Arapaho/Roosevelt National Forest. Travel est on I-70 about 2 miles past Idaho Springs and take exit 238, Fall River Road. Drive north on Fall River Road for 9 miles to the town of Alice. Turn left on Silver Creek then turn right on Texas Drive past the Glory Hole Mine. Take first dirt road on the right after the mine. It is a rough road. Hike around the west side of Loch Lomond and uphill beside the inlet stream about one mile to Reynolds Lake. Turn north and hike around the east side of Reynolds Lake and then 0.1 miles to Steuart Lake.
Size: 7 acres; 15 feet maximum depth.
Fish: Brook and lake trout.
Agency: Agriculture Ditch and Reservoir Company.
Comments: USGS Empire quad; elevation 11,400 feet. No boats.

38 **Reynolds Lake**
Location: Clear Creek County. Arapaho/Roosevelt National Forest. Travel west on I-70 about 2 miles past Idaho Springs and take exit 238, Fall River Road. Drive north on Fall River Road for 9 miles to the town of Alice. Turn left on Silver Creek then turn right on Texas Drive past the Glory Hole Mine. Take the first dirt road on the right after the mine. It is a rough road. Hike around the west side of Loch Lomond and uphill along the inlet stream about 0.1 miles to Reynolds Lake.
Size: 3 acres; 28 feet maximum depth.
Fish: Brook trout.
Agency: Agriculture Ditch and Reservoir Company.
Comments USGS Empire quad; elevation 11,200 feet. Non-motorized boats only

39 **Loch Lomond Lake**
Location: Clear Creek County. Arapaho/Roosevelt National Forest. Travel west on I-70 about 2 miles past Idaho Springs and take exit 238, Fall River Road. Drive north on Fall River Road for 9 miles to the town of Alice. Turn left on Silver Creek then turn right and take Texas Drive past the Glory Hole Mine. Take the first dirt road on the right after the mine. It is a rough road. Loch Lomond Lake is about 3.3 miles from Alice. **Note:** Roads leading to this lake are rough, unimproved dirt and/or gravel. Depending upon road conditions, a 4-wheel drive vehicle may be needed to reach the lake. Persons utilizing 2-wheel drive vehicles may need to park off the road and hike to the lake.
Size: 23 acres; 76 feet maximum depth.
Fish: Brook, brown, and lake trout.

Agency: Agriculture Ditch and Reservoir Company.
Comments: USGS Empire quad; elevation 11,180 feet.
Boats allowed. **Special Regulations:** The bag and possession limit for lake trout (Mackinaw) is one fish, 20 inches
or longer.

40 Lake Caroline

Location: Clear Creek County. Arapaho/Roosevelt
National Forest. Travel west on I-70 about 2 miles past
Idaho Springs and take exit 238, Fall River Road. Drive
north on Fall River Road for 9 miles to the town of Alice.
Turn left on Silver Creek, then turn right on Texas Drive
past the Glory Hole Mine. Take the first dirt road right
after the mine. It is a rough road. Drive to Loch Lomond.
Hike west cross country for a half mile to Lake Caroline.
Note: Roads leading to this lake are rough, unimproved
dirt and/or gravel. Depending upon road conditions, a 4-
wheel drive vehicle may be needed to reach the lake. Per-
sons utilizing 2-wheel drive vehicles may need to park off
the road and hike to the lake.
Size: 8.6 acres; 58 feet maximum depth.
Fish: Cutthroat trout.
Agency: USFS-Clear Creek Ranger District.
Comments: USGS Empire quad; elevation 11,840 feet.
Non-motorized boats only.

41 Saint Mary's Lake

Location: Clear Creek County. Arapaho/Roosevelt
National Forest. Travel west on I-70 about 2 miles past
Idaho Springs and take exit 238, Fall River Road. Drive 10
miles on Fall River Road (past the town of Alice) to St.
Mary's Glacier Lodge. Park and hike northwest about one-
quarter mile up an old road to Saint Mary's Lake.
Size: 7.2 acres; 21 feet maximum depth.
Fish: Brook trout.
Agency: USFS Clear Creek Ranger District.
Comments: USGS Empire quad; elevation 10,710 feet.
Non-motorized boats only.

42 Fall River Reservoir

Location. Clear Creek County. Arapaho/Roosevelt
National Forest. Travel west on I-70 about 2 miles past
Idaho Springs and take exit 238, Fall River Road. Drive
north on Fall River Road for 7.3 miles. Where the main
road turns uphill, take the narrow dirt road which follows
the Fall River. Follow this road for 2.3 miles to fork in the
road. Take right fork for 2 miles to reservoir. **Note:** Roads
leading to this lake are rough, unimproved dirt and/or
gravel. Depending upon road conditions, a 4-wheel drive
vehicle may be needed to reach the lake. Persons utilizing
2-wheel drive vehicles may need to park off the road and
hike to the lake.
Size: 17 acres; 80 feet maximum depth.
Fish: Rainbow and cutthroat trout.
Agency: Agriculture Ditch and Reservoir Company.
Comments: USGS Empire quad; elevation 10,880 feet.
Non-motorized boats only.

43 Slater Lake

Location: Clear Creek County. Arapaho/Roosevelt
National Forest. Travel west on I-70 about 2 miles past
Idaho Springs and take exit 238, Fall River Road. Drive
north on Fall River Road for 7.3 miles. Where the main
road turns uphill, take the narrow dirt road which follows
the Fall River. Follow this road for 2.3 miles to the fork in
the road. Take left fork 1.2 miles to Chinn's Lake. It is a
rough road. Stay on left side of the dam and go one mile to
Sherwin Lake. Then from the northwest side of Sherwin
Lake, hike following the stream one-quarter mile to the
lake. **Note:** Roads leading to this lake are rough, unimproved dirt and/or gravel. Depending upon road conditions,
a 4-wheel drive vehicle may he needed to reach the lake.
Persons utilizing 2-wheel drive vehicles may need to park
off the road and hike to the lake.
Size: 7.2 acres; 4.5 feet maximum depth.
Fish: Cutthroat trout.
Agency: USFS-Clear Creek Ranger District.
Comments: USGS Empire quad; elevation 11,440 feet.
Non-motorized boats only.

44 Sherwin Lake

Location: Clear Creek County. Arapaho/Roosevelt
National Forest. Travel west on I-70 about 2 miles past
Idaho Springs and take exit 238, Fall River Road. Drive
north on Fall River Road for 7.3 miles. Where the main
road turns uphill, take the narrow dirt road which follows
the Fall River. Follow this road for 2.3 miles to the fork in
the road. Take left fork 1.2 miles to Chinn's Lake. It is a
rough road. Stay on left side of dam and go 0.1 miles to
lake. **Note:** Roads leading to this lake are rough, unimproved dirt and/or gravel. Depending upon road conditions,
a 4-wheel drive vehicle may be needed to reach the lake.
Persons utilizing 2-wheel drive vehicles may need to park
off the road and hike to the lake.
Size: 8.6 acres; 21 feet maximum depth.
Fish: Brook, rainbow, and cutthroat trout.
Agency: USFS-Clear Creek Ranger District.
Comments: USGS Empire quad; elevation 11,090 feet.
Non-motorized boats only.

45 Chinn's Lake

Location: Clear Creek County. Arapaho/Roosevelt
National Forest. Travel west on I-70 about 2 miles past
Idaho Springs and take exit 238, Fall River Road. Drive
north on Fall River Road for 7.3 miles. Where the main
road turns uphill, take the narrow dirt road which follows
the Fall River. Follow this road for 2.3 miles to the fork in
the road. Take left fork 1.2 miles to Chinn's Lake. Note:
Roads leading to this lake are rough unimproved dirt
and/or gravel. Depending upon road conditions, a 4-wheel
drive vehicle may be needed to reach the lake. Persons utilizing 2-wheel drive vehicles may need to park off the road
and hike to the lake.
Size: 10 acres; 30 feet maximum depth.
Fish: Rainbow trout and splake.
Agency: USFS-Clear Creek Ranger District.
Comments: USGS Empire quad; elevation 11,000 feet.
Non-motorized boats only.

Metro and Mountain Lakes

46 Bill Moore Lake
Location: Clear Creek County. Arapaho/Roosevelt National Forest. From Empire, drive north on Empire Creek Road to the abandoned Conqueror Mine. Go north and northwest on the winding 4WD road. Total distance is about 6 miles to Bill Moore Lake from Empire.
Size: 7 acres; 3 feet maximum depth.
Fish: Cutthroat trout.
Agency: USFS-Clear Creek Ranger District.
Comments: USGS Empire quad; elevation 11,280 feet. Non-motorized boats only.

47 Byron Lake
Location: Clear Creek County. Arapaho/Roosevelt National Forest. From Empire drive north on Empire Creek Road to the abandoned Conqueror Mine. Go north and northwest on the winding 4WD road. The total distance is about 6 miles from Empire to Bill Moore Lake. From the northwest side of Bill Moore Lake, follow the inlet stream, taking the north fork and hike one mile cross country following the stream to Byron Lake. There is no established trail. Use of topo map is advised.
Size: 2.8 acres; 9 feet maximum depth.
Fish: Cutthroat trout.
Agency: USFS-Clear Creek Ranger District.
Comments: USGS Empire quad; elevation 12,100 feet. Non-motorized boats only.

48 Ethel Lake
Location: Clear Creek County. Arapaho/Roosevelt National Forest. From Empire, drive north on Empire Creek Road to the abandoned Conqueror Mine. Go north and northwest on the winding 4WD road. Total distance is about 6 miles from Empire to Bill Moore Lake. From the northwest side of Bill Moore Lake, follow the inlet stream taking the south fork and hike west 1.2 miles following the stream to Ethel Lake. There is no established trail. Use of topo map is advised.
Size: 5 acres; 65 feet maximum depth.
Fish: Cutthroat trout.

Agency: USFS-Clear Creek Ranger District.
Comments: USGS Empire quad; elevation 12,560 feet. Non-motorized boats only.

49 Cone Lake
Location: Clear Creek County. Arapaho/Roosevelt National Forest. From Empire, drive west on US 40. Follow the highway to the summit of Berthoud Pass. Hike 2 miles east on trail to Cone Lake.
Size: 3 acres; 8 feet maximum depth.
Fish: Cutthroat trout.
Agency: USFS-Clear Creek Ranger District.
Comments: USGS Berthoud Pass quad; elevation 11,600 feet. Steep hiking. Non-motorized boats only.

50 Urad Reservoir - Upper
Location: Clear Creek County. Arapaho/Roosevelt National Forest. From Empire go about 7 miles west on US 40 to Big Ben Picnic Ground. Go west from hairpin turn onto Henderson Mine Road. Go a short distance and then turn south. Drive 2 miles on the dirt road to Urad Reservoir.
Size: 31 acres; 48 feet maximum depth.
Fish: Brook, rainbow trout (catchable size stocked), and brown trout.
Agency: Amax, Inc.
Comments: USGS Gray's Peak quad; elevation 10,720 feet. Mizpaw Campground is 1.5 miles on US 40 from Big Ben Picnic Ground. No boats allowed.

51 Hassell Lake
Location: Clear Creek County. Arapaho/Roosevelt National Forest. From Empire go about 7 miles west on US 40 to Big Ben Picnic Ground. Go west from hairpin turn onto Henderson Mine Road. Go a short distance and then turn south. Drive 2 miles on the dirt road to Urad Reservoir. Hike a half mile northwest from the reservoir along tributary stream to Hassell Lake.
Size: 8.6 acres; 7 feet maximum depth.
Fish: Brook trout.
Agency: USFS-Clear Creek Ranger District.
Comments: USGS Gray's Peak quad; elevation 11,360 feet. Mizpaw Campground is 1.5 miles on US 40 from Big Ben Picnic Ground. Non-motorized boats only.

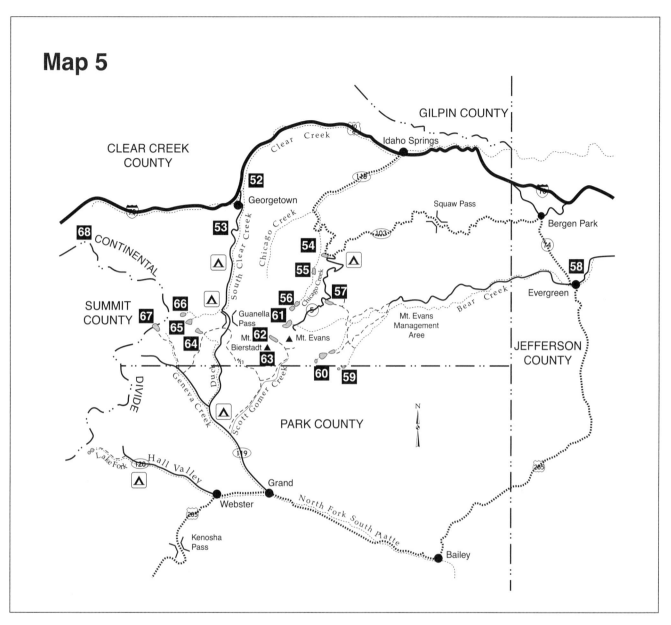

Map 5

52 **Georgetown Lake**
Location: Clear Creek County. Arapaho/Roosevelt National Forest. From Idaho Springs, continue west on I-70 to Georgetown exit. Go south to first stop sign, turn east and drive to the lake.
Size: 54.3 acres; 11 feet maximum depth.
Fish: Rainbow trout (catchable size stocked), brown and brook trout.
Agency: City of Georgetown.
Comments: USGS Georgetown quad; Elevation 8,460 feet. Clear Creek Campground is 4 miles south on Guanella Pass Road. Non motorized boats only. Handi-capped-accessible fishing pier. Bighorn sheep viewing. Ice fishing.

53 **Clear Lake**
Location: Clear Creek County. Arapaho/Roosevelt National Forest. From Georgetown, drive south on Guanella Pass Road about 4 miles. Clear Lake is just east of the road.
Size: 24 acres, 98 feet maximum depth.
Fish: Rainbow trout (catchable size stocked) and brook trout, sucker.
Agency: Public Service Company
Comments: USGS Idaho Springs quad; elevation 9,873 feet.Clear Lake Campground is about one mile south of the lake. No boats.

54 **Echo Lake**
Location: Clear Creek County. Arapaho/Roosevelt National Forest. From Idaho Springs, drive south 12 miles on Colorado 103. Lake is south of the highway just west of Mt. Evans Road.
Size: 18.2 acres; 7 feet maximum depth.

Metro and Mountain Lakes

Fish: Rainbow trout (catchable size stocked).
Agency: Denver Parks & Recreation.
Comments: USGS Idaho Springs quad; elevation 10,720 feet. West Chicago Creek Campground is 8 miles west. No boats.

55 Idaho Springs Reservoir
Location: Clear Creek County. Arapaho/Roosevelt National Forest. From Idaho Springs, drive south on Colorado 103 about 8 miles (past the Chicago Forks Picnic Grounds) to Chicago Creek Road. Drive south 1.5 miles and then hike 1.5 miles south on road to reservoir.
Size: 20 acres; 30 feet maximum depth.
Fish: Brook, rainbow and cutthroat trout, sucker.
Agency: City of Idaho Springs.
Comments: USGS Idaho Springs quad; elevation 10,600 feet. West Chicago Creek Campground is west. No boats.

56 Chicago Lakes
Location: Clear Creek County. Arapaho/Roosevelt National Forest. From Idaho Springs, drive south on Colorado 103 for about 8 miles (past the Chicago Forks Picnic Grounds) to Chicago Creek Road. Drive south 1.5 miles and then hike 1.5 miles south on road to Idaho Springs Reservoir. Hike around the west side of the reservoir and follow the creek upstream. Go 2 miles south to Chicago Lakes. Alternate route from Idaho Springs: Drive 12 miles south on Colorado 103 to Echo Lake. Hike west on Chicago Reservoir Trail to Chicago Creek Road, then continue as above.
Size: Upper Lake (10 acres; 41 feet maximum depth); Lower Lake (26 acres; 74 feet maximum depth).
Fish: Cutthroat trout, rainbow.
Agency: USFS-Clear Creek Ranger District.
Comments: USGS Idaho Springs and Mt. Evans quads; elevation 11,600 feet. Camping at West Chicago Creek Campground. Non-motorized boats only.

57 Lincoln Lake
Location: Clear Creek County. Arapaho/Roosevelt National Forest. From Idaho Springs drive 12 miles south on Colorado 103 to Echo Lake. Hike one mile east on Beaverdam Trail #46. Turn south and hike over 3 miles to next fork. Turn west and hike over a half mile on Trail #45 to Lincoln Lake.
Size: 12.8 acres; 61 feet maximum depth.
Fish: Brook and lake trout, sucker.
Agency: USFS-Clear Creek Ranger District.
Comments: USGS Harris Park quad; elevation 11,620 feet. Very difficult hike. The lake lies some 900 vertical feet directly below Mt. Evans Road, 3.5 miles south of Echo Lake junction. Non-motorized boats only.

58 Evergreen Lake
Location: Jefferson County. From Denver, go west on I-70 to the El Rancho exit. Go south on Colorado 74 to Evergreen. The lake is south of Colorado 74 and Upper Bear Creek Road.
Size: 42 acres; 22 feet maximum depth.

Fish: Brown and rainbow trout (catchable size stocked), sucker, tiger muskie and splake.
Agency: Evergreen Parks & Recreation District.
Comments: USGS Evergreen quad; elevation 7,072 feet. Boating by Evergreen permit only. No power boats. Open late May from 5 AM to 10 PM for fishing. Handicapped-accessible fishing pier and marsh viewing boardwalk. Parking available above and below dam.

59 Roosevelt Lakes
Location: Park County. Arapaho/Roosevelt National Forest. From Denver, drive south on US 285. Near the top of Crow Hill, 3 miles north of Bailey, turn northwest on Forest Road 100. Drive about 9 miles to Deer Creek Campground at the end of the road. Hike north 4 miles on Tanglewood Creek Trail #636 to Roosevelt Lakes. Alternate route from Evergreen: drive about 6 miles west on Upper Bear Creek Road. Take the west fork after Brookvale and continue west for 2 miles. Take the south fork and go 2 miles. Then take the north fork and continue west for 4 miles to Camp Rock Campground. Hike about 4.5 miles southwest on Beartrack Lakes Trail #43 to Beartrack Lakes. From here hike one mile southeast on Trail #78 to Roosevelt Lakes.
Size: 2 lakes; 6 acres total; 22 feet maximum depth.
Fish: Cutthroat, rainbow and brook trout.
Agency: USFS-Clear Creek Ranger District.
Comments: USGS Harris Park quad; elevation 10,400 feet. Non-motorized boats only.

60 Beartrack Lakes
Location: Clear Creek County. Arapaho/Roosevelt National Forest. From Denver, go west on I-70 to El Rancho exit. Go south on Colorado 74 to Evergreen. From Evergreen, drive about 6 miles west on Upper Bear Creek Road. Take the west fork after Brookvale and continue west for 2 miles. Take the south fork and go 2 miles. Then take the north fork and continue west for 4 miles to Camp Rock Campground. Hike about 4.5 miles southwest on Beartrack Lakes Trail #43 to the lakes.
Size: Upper Lake (5 acres; 25 feet maximum depth); Lower Lake (11 acres; 28 feet maximum depth).
Fish: Brook and cutthroat trout.
Agency: USFS-Clear Creek Ranger District.
Comments: USGS Harris Park quad; elevation 10,500 feet. Camping at Beartrack Lakes and Camp Rock Campground then follow trail signs. Dogs must be on 6-foot leash in elk management area. Non-motorized boats only.

61 Summit Lake
Location: Clear Creek County. Arapaho/Roosevelt National Forest. From Idaho Springs, drive 12 miles south on Colorado 103 to Echo Lake. Head east from lake and turn south on Mt. Evans Road, Colorado 5. Drive 9 miles, and the lake is just west of the road.
Size: 32.8 acres, 70 feet maximum depth.
Fish: Rainbow and cutthroat trout.
Agency: Denver Parks & Recreation.
Comments: USGS Mt. Evans quad; elevation 12,900 feet. Road-side fishery. Non-motorized boats only. Open 5 AM to 11 PM.

Metro and Mountain Lakes

62 Abyss Lake
Location: Clear Creek County. Pike National Forest. From Grant, drive north on Guanella Pass Road (Forest Road 118) to Burning Bear Campground. Hike 3.5 miles northwest on Scott Gomer Creek Trail to Lake Fork Trail #602, just past Deer Creek Trail #603. Hike northwest 3 miles on Lake Fork Trail to Abyss Lake. Alternate route from Georgetown: Drive south on Guanella Pass Road. Then travel about 6 miles south from the summit of the pass to Burning Bear Campground and continue as above. Abyss Lake is between Mt. Evans and Mt. Bierstadt.
Size: 18 acres; 50 feet maximum depth.
Fish: Rainbow and cutthroat trout.
Agency: USFS-South Platte Ranger District.
Comments: USGS Mt. Evans quad; elevation 12,640 feet. Frozen over until mid-June or later. The lake is in a harsh environment. Camping at Burning Bear and Geneva Creek Campgrounds. Non-motorized boats only.

63 Frozen Lake
Location: Clear Creek County. Pike National Forest. From Grant, drive north on Guanella Pass Road (Forest Road 118) to Burning Bear Campground. Hike 3.5 miles northwest on Scott Gomer Creek Trail to Lake Fork Trail. Pass two trails going east and continue one mile north on Scott Gomer Creek Trail. Follow the creek north where the trail heads west. Hike 2 miles upstream to Frozen Lake. The hike is over rugged terrain. Use of a topo map is advised. Alternate route from Georgetown: Drive south on Guanella Pass Road. Then travel about 6 miles south from the summit of the pass to Burning Bear Campground then continue as above.
Size: 7 acres; 33 feet maximum depth.
Fish: Cutthroat trout.
Agency: USFS-South Platte Ranger District.
Comments: USGS Mt. Evans quad; elevation 12,960 feet. Harsh environment. Lake is frozen over until mid-June or later. Mt. Bierstadt at 14,060 feet elevation looms over the lake. Camping at Burning Bear and Geneva Creek Campgrounds. No boats.

64 Square Top Lakes
Location: Clear Creek Arapaho/Roosevelt National Forest. From Georgetown, drive south on Guanella Pass Road to the summit of the pass. Hike 2 miles west on the trail to Square Top Lakes.
Size: 2 lakes; 10 acres each.
Fish: Cutthroat trout.
Agency: USFS-Clear Creek Ranger District.
Comments: USGS Mt. Evans quad; elevation 12,160 feet. Guanella Pass and Clear Lake Campgrounds nearby on the south fork of Clear Creek. Non-motorized boats only.

65 Silver Dollar Lake
Location: Clear Creek County Arapaho/Roosevelt National Forest. From Georgetown, go 7 miles south on Guanella Pass Road to Guanella Pass Campground. Just past the campground take the west fork. Drive one mile to Naylor Lake, which is private property. Hike west along the south side of Naylor Lake to a trail along the creek. Hike a half mile west on the west fork of the trail to Silver Dollar Lake. Alternate route from Grant: drive north on Forest Road 118, Guanella Pass Road, over the pass to Guanella Pass Campground. Continue from campground as in above directions.
Size: 18.6 acres; 73 feet maximum depth.
Fish: Cutthroat trout.
Agency: USFS-Clear Creek Ranger District.
Comments: USGS Mt. Evans and Montezuma quads; elevation 11,950 feet. Non-motorized boats only.

66 Murray Lake
Location: Clear Creek County Arapaho/Roosevelt National Forest. From Georgetown, go 7 miles south on Guanella Pass Road to Guanella Pass Campground. Just past the campground, take the west fork. Drive one mile to Naylor Lake, which is private property. Hike west along the south side of Naylor Lake to a trail along the creek. Hike one mile northwest on the north fork of the trail to Murray Lake.
Size: 11.4 acres; 38 feet maximum depth.
Fish: Cutthroat trout.
Agency: USFS-Clear Creek Ranger District.
Comments: USGS Mt. Evans and Montezuma quads; elevation 12,080 feet. Non-motorized boats only.

67 Shelf Lake
Location Clear Creek County. Pike National Forest. From Grant, drive north on Guanella Pass Road, Forest Road 118, to Duck Creek Picnic Ground. Take the west fork of the road and drive 3 miles northwest up Geneva Creek Road, Forest Road 119, to Smelter Gulch. Hike 3 miles up the trail to Shelf Lake.
Size: 9.5 acres; 40 feet maximum depth.
Fish: Cutthroat trout.
Agency: USFS-South Platte Ranger District.
Comments: USGS Montezuma quad; elevation 12,000 feet. Harsh environment. Lake is frozen over until mid-June or later. Mt. Bierstadt at 14,060 feet elevation looms over the lake. Camping at Burning Bear and Geneva Creek Campgrounds. Non-motorized boats only.

68 Gibson Lake
Location: Park County. Pike National Forest. From Denver, drive south on US 285 to Grant. From Grant, drive west 3 miles on US 285 and turn north at Webster. Drive 7 miles northwest on Hall Valley Road, Forest Road 120, to trailhead at Gibson Lake Trail Picnic Ground. Hike 2.5 miles south, then west on Gibson Lake Trail #633.
Size: 3 acres; 23 feet maximum depth.
Fish: Brook trout.
Agency: USFS-South Platte Ranger District
Comments: USGS Jefferson quad; 11,500 feet. Hall Valley Campground near trailhead. Non-motorized boats only.

Appendix 1

Stocked
Lakes and Reservoirs
of
Eastern Colorado

Note: For easy reference the state has been divided into two parts, eastern and western sections. A reference point (106°) for the eastern and western sections would be at Lake Dillon at I-70 in Summit County. An imaginary line (106°) running north and south from this point divides the listed lakes into east and west sections.

DOW Stocked Lakes Code Explanation

Major River Drainage Code

AR	ARKANSAS RIVER
CR	COLORADO RIVER
GR	GREEN RIVER
GU	GUNNISON RIVER
NP	NORTH PLATTE RIVER
RE	REPUBLICAN RIVER
RG	RIO GRANDE RIVER
SJ	SAN JUAN RIVER
SP	SOUTH PLATTE RIVER
WR	WHITE RIVER
YP	YAMPA RIVER

County Codes

ADA	ADAMS		GUN	GUNNISON
ALA	ALAMOSA		HIN	HINSDALE
ARA	ARAPAHOE		HUE	HUERFANO
ARC	ARCHULETA		JAC	JACKSON
BAC	BACA		JEF	JEFFERSON
BEN	BENT		KIO	KIOWA
BOU	BOULDER		KIT	KIT CARSON
CHA	CHAFFEE		LAK	LAKE
CLE	CHEYENNE		LAP	LA PLATA
CON	CONEJOS		LAR	LARIMER
COS	COSTILLA		LAS	LAS ANIMAS
CRO	CROWLEY		LIN	LINCOLN
CUS	CUSTER		LOG	LOGAN
DEL	DELTA		MES	MESA
DEN	DENVER		MIN	MINERAL
DOL	DOLORES		MOR	MOFFAT
DOU	DOUGLAS		MTZ	MONTEZUMA
EAG	EAGLE		MON	MONTROSE
ELB	ELBERT		MOR	MORGAN
ELP	EL PASO		OTE	OTERO
FRE	FREMONT		OUR	OURAY
GAR	GARFIELD		PAR	PARK
GIL	GILPIN		PHI	PHILLIPS
GRA	GRAND		PIT	PITKIN
			PRO	PROWERS
			PUE	PUEBLO
			RBL	RIO BLANCO
			RGR	RIO GRANDE
			ROU	ROUTT
			SAG	SAGUACHE
			SNJ	SAN JUAN
			SNM	SAN MIGUEL
			SED	SEDGWICK
			SUM	SUMMIT
			TEL	TELLER
			WAS	WASHINGTON
			WEL	WELD
			YUM	YUMA

Fish Species Codes

BCF	CATFISH — BLUE
BCR	BLACK CRAPPIE
BGL	BLUEGILL
BHD	BULLHEAD
BRK	BROOK TROUT
CCF	CATFISH — CHANNEL
CRA	CRAPPIE
CUT	CUTTHROAT
FLC	CATFISH — FLATHEAD
GOL	GOLDEN TROUT
GRA	GRAYLING
KOK	KOKANEE SALMON
LMB	BASS — LARGE MOUTH
LOC	BROWN TROUT
LXB	TIGER TROUT
MAC	MACKINAW (LAKE) TROUT
NPK	NORTHERN PIKE
RBT	RAINBOW TROUT
SAG	SAUGEYE
SBS	STRIPED BASS
SGR	SAUGER
SMB	BASS — SMALL MOUTH
SPD	BASS — SPOTTED
SPL	SPLAKE
SQF	SQUAWFISH
SXW	WIPER
TGM	TIGER MUSKIE
WAL	WALLEYE
WBA	WHITE BASS
WHF	WHITE FISH
YPE	YELLOW PERCH

Heading Explanation

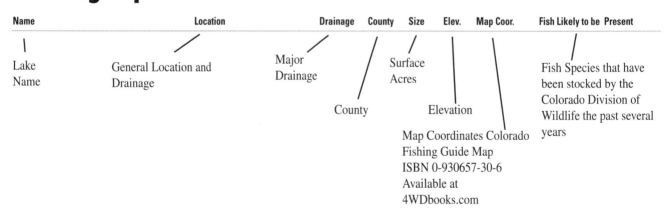

Name	Location	Drainage	County	Size	Elev.	Map Coor.	Fish Likely to be Present
Lake Name	General Location and Drainage	Major Drainage	County	Surface Acres	Elevation	Map Coordinates Colorado Fishing Guide Map ISBN 0-930657-30-6 Available at 4WDbooks.com	Fish Species that have been stocked by the Colorado Division of Wildlife the past several years

Name	Location	Drainage	County	Size	Elev.	Map Coor.	Fish Likely to be Present
A							
ABYSS LAKE	FK SCOTT GOMER CR, Mt Evans	SP	CLE	18.0	12650	C 7	RBT, CUT
ADOBE CREEK RES.	N of Las Animas	AR	BEN	5147.0	4128	F 12	LMB, CCF, BCF, WAL, SAG, TGM
AGNES LAKE, LOWER	MICHIGAN R, S of Cameron Pass	NP	JAC	22.0	10663	B 7	RBT, CUT
AGNES LAKE, UPPER	MICHIGAN R, S of Cameron Pass	NP	JAC	3.0	11160	B 7	RBT, CUT
ANTERO RESERVOIR	S FK S PLATTE R, NE/Antero Jct	SP	PAR	4102.0	8940	E 7	RBT, CUT, LOC, SPL
ANTICLINE LAKE	1/3 mi below Pueblo Res dam	AR	PUE	10.0	4750	F 9	SPL, RBT, CCF, BGL
ARAPAHO LAKE #2	ARAPAHO CR, S FK COLO. R	CR	GRA	1.5	11 240	B 7	CUT
ARAPAHO LAKE #2	S BOULDER CR, NW/East Portal	SP	GIL	2.5	11180	C 7	CUT
ARAPAHO LAKE #3	S BOULDER CR, NW/East Portal	SP	GIL	10.0	11140	C 7	CUT
AURORA RESERVOIR	4 mi E of Aurora, on Quincy Ave.	SP	ARA	820.0	5710	C 9	RBT, WAL, SXW, BCF, CCF
B							
BALMAN RESERVOIR	LAKE CR, W of Hillside	AR	CUS	5.0	9440	F 7	RBT, CUT, GRA
BANANA LAKE	NW of Eldora, up strm/Diamond L	SP	BOU	1.5	11320	C 7	CUT, LOC
BANJO LAKE	MID BRUSH CR, NW of Westcliffe	AR	CUS	5.4	12350	F 7	CUT
BARBOUR POND #1	N of I-25 exit 240, Del Camino	SP	WEL	48.0	4830	B 9	RBT, CCF, BGL
BARBOUR POND #2	N of I-25 exit 240, Del Camino	SP	WEL	25.0	4830	B 9	RBT, CCF, BGL
BARBOUR POND #3	N of I-25 exit 240, Del Camino	SP	WEL	10.0	4830	B 9	RBT, CCF, BGL
BARKER RESERVOIR	MID BOULDER CR, E/Nederland	SP	BOU	380.0	8186	C 7	RBT, LOC
BARNES MEADOW RES	JOE WRIGHT CR, E/Chambers Lk	SP	LAR	112.8	9100	A 7	RBT, CUT
BARR LAKE	SE of Brighton	SP	ADA	1660.0	5095	C 9	BCF, CCF, SXW, FLC, BCF, RBT, WAL
BEAR LAKE	CUCHARAS CR, SW of La Veta	AR	HUE	2.5	10500	H 8	RBT
BEAR LAKE, LOWER	CUCHARAS CR, SW of La Veta	AR	HUE	3.0	10460	H 8	RBT
BEARTRACK LK, LOWER	BEARTRACK CR, SE of Mt Evans	SP	CLE	11.4	11160	C 7	SPL
BEARTRACK LK. UPPER	BEARTRACK CR, SE of Mt Evans	SP	CLE	4.5	11950	C 7	CUT
BECKWITH RESERVOIR	GREENHORN CR, Cob City	AR	PUE	60.0	6070	G 9	CCF, RBT
BELLAIRE LAKE	unnmd to S LONE PINE, NW/Rustic	SP	LAR	9.8	8640	A 7	RBT
BETTY LAKE	S FK BOULDER CR, W of Eldora	SP	BOU	8.5	11440	C 7	CUT
BIG MAC POND	McCelland SWA, E/Rocky Ford	AR	OTE	1.0	4150	F 11	HGB, CCF, BCR, LMB
BIG THOMPSON POND #1	SE/Loveland, SW/I-25 exit 257	SP	LAR	3.5	4850	B 8	CCF, BGL, LMB
BIG THOMPSON POND #2	SE/Loveland, SW/I-25 exit 257	SP	LAR	6.1	4850	B 8	CCF, BGL, LMB
BIG THOMPSON POND #3	SE/Loveland, SW/I-25 exit 257	SP	LAR	3.0	4850	B 8	CCF, BGL, LMB
BIG THOMPSON POND #4	SE/Loveland, SW/I-25 exit 257	SP	LAR	0.5	4850	B 8	LMB, CCF. BGL
BIG THOMPSON POND #5	SE/Loveland, SW/I-25 exit 257	SP	LAR	8.0	4850	B 8	CCF, BGL, LMB
BITTERSWEET LAKE	W of Greeley	SP	WEL	3.0	4810	B 9	RBT
BLACK HOLE AT TWO BUTTES	NE/Springfield, below dam	AR	BAC	1.0	4000	G 13	RBT
BLANCA WILDLIFE AREA PDS	DRY LAKES, NE of Alamosa (BLM)	RG	ALA	230.0	7520	G 7	RBT
BLIND LAKE, UPPER	RITO ALTO CR, W/Rito Alto Lake	RG	SAG	7.9	11920	F 7	CUT
BLUE LAKE	BOWEN GULCH, NW/Grand Lake	CR	GRA	2.0	10690	B 7	BRK, CUT
BLUE LAKE	FALL CR, NW of Chambers Lake	SP	LAR	17.2	10720	A 7	CUT, RBT
BLUE LAKE	S ST VRAIN, NW of Ward	SP	B0U	25.7	11320	B 7	CUT, BRK
BLUE LAKE	CUCHARAS CR, SW of La Veta	AR	HUE	1.0	10400	H 8	CUT, RBT
BLUE LAKE	HOLBROOK CR, W side/Mt Blanca	RG	ALA	4.9	12140	G 7	CUT
BLUE LAKE, LOWER	TURKEY CR, W of Rye	AR	PUE	1.2	11280	G 8	RBT, CUT
BLUE LAKE, UPPER	S ST VRAIN, NW of Ward	SP	BOU	7.2	11833	B 7	CUT, BRK
BOB LAKE	S FK BOULDER CR, W of Eldora	SP	BOU	8.5	11640	C 7	CUT
BOEDECKER RES	SW of Loveland, N/Lon Hagler R	SP	LAR	373.0	6062	B 8	WAL, LMB, CCF, BGL
BONNY PONDS WEST	W of Bonny Reservoir	AR	YUM	6.0	3690	C 14	BGL, CCF
BONNY RESERVOIR	N of Burlington, State Park	RE	YUM	1924.0	3670	C 14	WAL, SXW, CCF, LMB, SXW, YPE
BOULDER RESERVOIR	N of Boulder	NE	BOU	660.0	5600	B 8	WAL, RBT, LMB, CCF
BOWEN LAKE	BOWEN GULCH, NW of Grand Lake	CR	GRA	10.0	11019	B 7	CUT
BOX ELDER LAKE #3	at Wellington, aka Smith Lk	SP	LAR	30.0	5257	A 8	SXW, CCF, BGL
BOYD LAKE (STATE PARK)	NE of Loveland, I-25 exit 257	SP	LAR	1674.0	4958	B 8	RBT, LOC, BCR, WAL, CCF, WBA
BOYD ROSE POND	S of Pritchet	AR	BAC	1.0	4800	H 13	
BRAINARD LAKE	S ST VRAIN, W of Ward	SP	BOU	15.6	10360	B 7	RBT, LOC, BRK
BRENT LAKE	Cameron Pass	SP	LAR	5.7	11480	A 7	CUT
BRUSH CR LAKE, LOW.	N BRUSH CR, NW of Westcliffe	AR	CUS	45.6	11400	F 7	CUT, BRK
BRUSH CR LAKE, UPP.	N BRUSH CR, NW of Westcliffe	AR	CUS	44.6	11560	F 7	CUT, BRK
BRUSH HOLLOW RES.	NW of Penrose	AR	FRE	186.0	5500	F 8	WAL, BGL, CRA, CCF, RBT
BUSHNELL LAKE, LOW.	STOUT CR, SW of Howard	AR	FRE	4.0	11120	F 7	CUT
BUSHNELL LAKE, UPP.	STOUT CR, SW of Howard	AR	FRE	8.5	11900	F 7	CUT
BUTTON ROCK RES	N ST VRAIN, W of Lyons	SP	BOU	120.0	6400	B 8	RBT, SPL
BYRON LAKE	MILL CR, NW of Bill Moore Lk	SP	CLE	2.8	12050	C 7	CUT
C							
CAREY LAKES	W BRANCH LARAMIE, NW/Chambers L	NP	LAR	14.5	11303	A 7	CUT
CARIBOU LAKE	ARAPAHO CR, SE of Lk Granby	CR	GRA	7.0	11147	B 8	CUT
CAROLINE LAKE	FALL R, N of Empire	SP	CLE	8.6	11920	C 7	CUT
CARTER RESERVOIR	W of Berthoud	SP	LAR	1444.9	5781	B 8	RBT, SPL, MAC, KOK, LMB
CASTLE ROCK POND #1	CASTLE ROCK GUL, TROUT CR	AR	CHA	20.0	9100	E 7	CUT
CATAMOUNT RES, NORTH	CATAMOUNT CR, S/Woodland Park	AR	TEL	210.0	9344	E 8	CUT
CATAMOUNT RES, SOUTH	S CATAMOUNT CR, S/Woodland Park	AR	TEL	140.0	9200	E 8	CUT
CHAMBERS LAKE	LARAMIE R, NE of Cameron Pass	SP	LAR	254.5	9153	A 7	RBT, MAC, SPL
CHANCELLOR PONDS	PERLY-DOSS CANYONS, PURGATOIRE	AR	LAS	4.0	5200	H 11	BGL, CCF, LMB
CHARTIERS POND	N of Brush	SP	MOR	5.0	4230	B 11	LMB
CHATFIELD POND #1	Big Pd S of Chatfield Res area	SP	JEF	54.6	5430	C 8	LMB, CCF
CHATFIELD POND #3	Small pond S of POND #1	SP	JEF	19.8	5430	C 8	LMB
CHATFIELD POND #4	SE of POND #1, across river	SP	DOU	12.0	5430	C 8	BGL
CHATFIELD POND #5	S of POND #4	SP	DOU	42.7	5430	C 8	BGL
CHATFIELD RESERVOIR	C470 & Wadsworth, S/Littleton, State P	SP	DOU	1087.0	5430	C 8	RBT, WAL, CUT BCF, SMB, CCF
CHEESMAN RESERVOIR	SW of Deckers, SP R #5	SP	DOU	875.1	7382	D 8	KOK, SPL, SMB, RBT
CHERRY CREEK RESERVOIR	I-225 & Parker Rd, S/Aurora, State Park	SP	ARA	889.0	5555	C 9	TGM, RBT, SXW, BCF, CCF, FLC, WAL, SMB
CHERRY LAKE	WILD CHERRY CR, SAN LUIS CR	RG	SAG	9.9	11769	F 7	CUT

Eastern Colorado Lakes

Name	Location	Drainage	County	Size	Elev.	Map Coor.	Fish Likely to be Present
CHICAGO LAKE #3	CHICAGO CR, N of Mt Evans	SP	CLE	1.0	11400	C 7	CUT
CHICAGO LAKE, LOWER	CHICAGO CR, N of Mt Evans	SP	CLE	26.0	11420	C 7	CUT
CHICAGO LAKE, UPPER	CHICAGO CR, N of Mt Evans	SP	CLE	10.0	11760	C 7	CUT
CHIHUAHUA LAKE	CHIHUAHUA GUL, PERU CR, E/Kyst	CR	SUM	6.2	11340	C 7	CUT
CHINNS LAKE	FALL R, W of Alice	SP	CLE	10.0	11035	C 7	RBT, BRK, LOC, CUT
CLAYTON LAKE	S BOULDER CR, SW of East Portal	SP	GIL	5.0	11000	C 7	CUT
CLEAR LAKE.	CLEAR CR, NE of Meadow Cr Res	NP	JAC	10.0	10580	A 7	CUT
CLEAR LAKE	S CLEAR CR, S of Georgetown	SP	CLE	24.0	9873	C 7	CUT
COLONY LAKE, NORTH #1	N COLONY CR, SW of Westclitfe	AR	CUS	5.3	11571	G 7	CUT, RBT
COLONY LAKE, NORTH #3	N COLONY CR, SW of Westclffe	AR	CUS	5.4	11730	G 7	CUT, RBT
COLONY LAKE. NORTH #4	N COLONY CR, SW of Westclitfe	AR	CUS	2.0	11600	G 7	CUT, RBT
COLONY LAKE, NORTH #5	N COLONY CR, SW of Westcliffe	AR	CUS	5.4	12485	G 7	CUT, RBT
COLONY LAKE. SOUTH. LOWER	S COLONY CR, SW o f Westcliffe	AR	CUS	6.0	11660	G 7	CUT, RBT
COLONY LAKE, SOUTH, UPPER	S COLONY CR, SW of Westclitfe	AR	CUS	16.3	12030	G 7	CUT, RBT
COLORADO SPRINGS RES #5	BOEHMER CR, SW of Colo Springs	SP	ELP	110.0	10900	E 8	CUT
COLUMBINE LAKE	MEADOW CR, FRASER R, SE/Granby	CR	GRA	6.3	11060	B 8	CUT
COMANCHE LAKE	BEAVER CR, S of Idylwilde	SP	LAR	8.4	10000	A 7	CUT, RBT
COMANCHE LAKE	HILTMAN CR, SW of Westclffe	AR	CUS	20.5	11665	F 7	CUT
COMANCHE POND	COMANCHE NATIONAL GRASSLAND	AR	BAC	20+	4800	H 13	CCF, BGL, LMB
CONE LAKE	BLUE CR, E of Berthoud Pass	SP	CLE	3.0	11600	C 7	CUT
CONEY LAKE, LOWER	CONEY CR, NW of Ward	SP	BOU	9.0	10600	B 8	CUT
CONEY LAKE	UPPER CONEY CR, NW of Ward	SP	BOU	16.0	10940	B 8	CUT
COPELAND LAKE	N ST VRAIN, N of Allenspark	SP	BOU	5.7	8312	B 8	RBT
CORONA LAKE	RANCH CR, E of Fraser	CR	GRA	16.6	11206	C 7	CUT
COTTON LAKE	COTTON CR, SAN LUIS CR	RG	SAG	9.9	11520	F 7	CUT
COTTONWOOD LAKE	COTTONWOOD CR, SAN LUIS CR	RG	SAG	3.0	12310	G 7	CUT, BRK
COTTONWOOD LAKE, SOUTH	COTTONWOOD CR, SAN LUIS CR	RG	SAG	2.0	11800	G 7	CUT, BRK
CRATER LAKE	HOLBROOK CR HDWTRS on Mt Blanc	RG	ALA	9.9	12700	G 7	CUT
CRATER LK #1, UPPER	S BOULDER CR, SW of East Portal	SP	GIL	8.5	11020	C 7	CUT
CRATER LK #2, MID NORTH	S BOULDER CR, SW of East Portal	SP	GIL	14.0	10600	C 7	SPL
CRATER LK #3, MID SOUTH	S BOULDER CR, SW of East Portal	SP	GIL	8.6	10600	C 7	SPL
CRAWFORD LAKE	HELLS CANYON CR, E/Lk Granby	CR	GRA	4.0	10130	B 8	CUT
CREEDMORE LAKE	S LONE PINE, Red Feather Group	SP	LAR	10.0	8320	A 7	BRK, LOC
CRESTONE LAKE, NORTH	N CRESTONE, NE of Crestone	RG	SAG	31.6	11560	F 7	CUT, BRK
CRESTONE LAKE, SOUTH	S CRESTONE, NE of Crestone	RG	SAG	8.9	11780	F 7	CUT, BRK
CRYSTAL CR RESERVOIR	S of Green Mountain Falls	AR	ELP	136.0	9222	E 8	CUT
D							
DAIGRE RESERVOIR	CUCHARAS R, NE of La Veta	AR	HUE	10.0	6960	G 9	CUT, SPL, RBT
DAVIS POND, JIM	TOBE CR, CHACUACO CR	AR	LAS	2.0	5600	H 11	BGL, CCF, LMB
DE WEESE RESERVOIR	GRAPE CR #2, N of Westcliffe	AR	CUS	208.0	7665	F 8	RBT, SPL, CUT LOC, SMB
DEADMAN LAKE	MID FK RANCH CR, FRASER R	CR	GRA	2.4	11261	C 7	CUT
DEADMAN LAKE #1 (LOWER)	DEADMAN CR, SAN LUIS CR	RG	SAG	2.5	11780	G 7	CUT, BRK
DEADMAN LAKE #2 (UPPER)	DEADMAN CR, SAN LUIS CR	RG	SAG	13.8	11704	G 7	CUT, BRK
DEADMAN LAKE, WEST	DEADMAN CR, SAN LUIS CR	RG	SAG	3.0	11765	G 7	CUT, BRK
DEEP LAKE	N FK MID BOULDER, W/Diamond Lk	SP	BOU	4.9	11540	C 7	CUT, RBT
DEPOORTER LAKE	JULESBURG	SP	SED	7.0	3477	A 14	BGL, LMB, RBT
DEVILS THUMB LAKE	JASPER CR, NW of Hessie	SP	BOU	11.5	11160	C 7	CUT
DIAMOND LAKE, LOW.	N FK MID BOULDER, NW of Hessie	SP	BOU	14.0	10920	C 7	SPL, RBT, BRK
DIAMOND LAKE, UPPER	N FK MID BOULDER, NW of Hessie	SP	BOU	6.0	11720	C 7	CUT, RBT, BRK
DIXON RES	E of mid dam Horsetooth Res	SP	LAR	50.0	5190	A 8	CCF, BGL
DOME ROCK PONDS	Dome Rock SWA	AR	TEL	7.0	8240	E 8	CUT
DOROTHY LAKE	N FK MID BOULDER, SW of Ward	SP	BOU	16.0	12061	B 8	CUT
DOUGLASS RESERVOIR	DRY CR. POUDRE R, W/Wellington	SP	LAR	565.0	5150	A 8	WAL, SXW, RBT, BCR, CCF, YPE, LMB
DOWDY LAKE	S LONE PINE, Red Feather Group	SP	LAR	115.1	8135	A 7	RBT, LOC, CUT, BRK
DRY LAKE, LOWER	DRY CR, SW of Westcliffe	AR	CUS	3.7	11820	F 7	CUT
DRY LAKE, MIDDLE	DRY CR, SW of Westcliffe	AR	CUS	6.8	11860	F 7	CUT
DRY LAKE, UPPER	DRY CR, SW of Westcliffe	AR	CUS	3.5	11940	F 7	CUT
E							
EAST PORTAL RESERVOIR	Alva Adams Tunnel, SW/Marys Lk	SP	LAR	5.0	8200	B 7	RBT
ECHO LAKE	CHICAGO CR, SW of Idaho Spgs	SP	CLE	18.2	10597	C 7	RBT, CUT
ECHO LAKE., LITTLE	MAMMOTH GULCH, SW of Tolland	SP	GIL	13.0	11185	C 7	RBT, CUT
ELEVEN MILE RES	S PLATTE R #7A, SW/Lake George	SP	PAR	3405.0	8620	E 7	RBT, KOK, LOC, CUT, MAC, NPK
EMMALINE LAKE	S FK POUDRE, CSU Pingre Campus	SP	LAR	5.7	10960	A 7	CUT
ESTES LAKE	BIG THOMPSON R #4. Estes Park	SP	LAR	185.0	7468	B 7	RBT, TGM
ETHEL LAKE	MILL CR, NW of Empire	SP	CLE	5.0	12600	C 7	CUT
EUREKA LAKE	M TAYLOR CR, W of Westcliffe	AR	CUS	18.1	11960	F 7	CUT
EVELYN LAKE	KINNEY CR, WMS FK COLO R	CR	GRA	2.0	11160	C 7	CUT
EVERGREEN RESERVOIR	BEAR CR #2, Town of Evergreen	SP	JEF	42.0	7072	C 8	RBT, TGM
F							
FALL RIVER RESERVOIR	FALL R, W of Alice	SP	CLE	24.7	10840	C 7	CUT
FITZLER POND	near Springfield	AR	LAS	2.0	4350	H 13	HGC, BGL, LMB
FLAGLER RESERVOIR	E of Flagler, N of I~70	RE	KIT	156.0	4707	D 13	WAL, BGL, SXW, CCF, BCR, YPE
FLATIRON RESERVOIR	DRY CR, NW of Carter Lk	SP	LAR	47.0	5473	B 8	RBT
FOREST LAKE #3.	UPPER S BOULDER CR, NW of East Portal	SP	GIL	4.3	10840	C 7	SPL, BRK, LOC
FOUNTAIN LAKE	E of Pueblo	AR	PUE	5.0	4700	F 9	RBT, CCF
FRANK EASEMENT POND	BIG THOMPSON R, S of Loveland	SP	WEL	20+	4900	B 8	LMB, CCF
FROZEN LAKE	LK FK SCOTT GOMER CR, N/Grant	SP	CLE	7.0	12960	C 7	CUT
G							
GEORGETOWN LAKE	CLEAR CR #2, at Georgetown	SP	CLE	54.3	8471	C 7	RBT
GOLDEN GATE STATE P. LKS	NOTT CR. (ALL)	SP	GIL	20.0	8300	C 8	RBT, BRK
GOODWIN LAKE, LOWER	GOODWIN CR, SW of Westcliffe	AR	CUS	2.5	11390	F 7	CUT
GOODWIN LAKE, UPPER	GOODWIN CR, SW of Westcliffe	AR	CUS	8.1	11580	F 7	CUT
GOURD LAKE	BUCHANAN CR. E of Lk Granby	CR	GRA	1.4	10840	B 7	CUT

Name	Location	Drainage	County	Size	Elev.	Map Coor.	Fish Likely to be Present
GRANBY RESERVOIR	COLO R #10, NE of Granby	CR	GRA	7280.0	8280	B 7	RBT, KOK, LOC, BRK, CUT, MAC
GRAND LAKE	City of Grand Lake	CR	GRA	506.0	8367	B 7	RBT, KOK, LOC, BRK, CUT, MAC
GRASS LAKE	POUDRE R #5, N of Peterson Lk	SP	LAR	5.7	9964	A 7	CUT
GRAYS LAKES	PERU CR, SNAKE R	CR	SUM	2.0	12440	C 7	CUT
GRAYS LAKES							
GREEN MTN. FALLS LAKE	Green Mountain Falls	AR	ELP	4.0	7600	E 8	RBT
GROSS RESERVOIR	S BOULDER CR, NE of Pinecliffe	SP	BOU	412.4	7287	C 8	RBT, KOK, SPL, MAC, TGM
H							
HALE PONDS	E of Bonny Res below dam	RE	YUM	15.0	3520	C 14	RBT, BGL, CCF, LMB, TPE, CRA
HANG LAKE	FALL CR, W of Blue Lk, Rawahs	NP	LAR	4.3	11120	A 7	CUT
HASTY LAKE	below John Martin Res dam	AR	BEN	73.0	3747	F 13	RBT, BGL, CCF, SXW
HEART LAKE	S BOULDER CR, SW of East Portal	SP	GIL	17.0	12218	C 7	CUT, RBT
HENRY LAKE	NE OF ORDWAY	AR	CRO	1120.0	4360	F 11	SXW, CCF, BCE LMB
HERMIT LAKE	M TAYLOR CR, W of Westcliffe	AR	CUS	20.5	11320	F 7	CUT, BRK
HOHNHOLZ LK #l, EAST	LARAMIE R, NW of Glendevey	NP	LAR	8.0	7880	A 7	RBT
HOHNHOLZ LK #2, LITTLE	LARAMIE R, NW of Glendevey	NP	LAR	8.0	7880	A 7	RBT, CUT
HOHNHOLZ LK #3, BIG	LARAMIE R, NW of Glendevey	NP	LAR	8.0	7880	A 7	LOC, CUT
HOLBROOK LAKE	E of Rocky Ford	AR	OTE	673.0	4150	F 11	RBT, SXW, CCF, LMB, BGL, CRA
HOLTORE LAKE #1, LOWER PD	N of Akron	SP	WAS	14.3	4250	B 12	CCF, BGL
HOLTORE LAKE #2, UPPER PD	N of Akron	SP	WAS	5.0	4255	B 12	CCF, BGL, LMB
HORN LAKE	HORN CR, SW of Westcliffe	AR	CUS	24.2	11830	F 7	CUT
HORN LK, NORTH (LITTLE)	HORN CR, SW of Westcliffe	AR	CUS	2.9	11632	F 7	CUT
HORSE CR RES /TIMBER CR)	HORSE CREEK, NE of Rocky Ford	AR	BEN	2900	4128	F 12	SAG, SXW, CCF
HORSESHOE LAKE	KINNEY CR, WMS FK COLO R	CR	GRA	9.2	11230	C 7	CUT
HORSESHOE LAKE	M TAYLOR CR, W of Westcliffe	AR	CUS	13.6	11960	F 7	CUT
HORSESHOE RESERVOIR	NE/Loveland, NW/I-25 exit 257	SP	LAR	652.0	4973	B 8	CCF, RBT, WAL, BGL
HORSESHOE RESERVOIR	W of Walsenburg, Lathrop State Park	AR	HUE	170.0	6450	G 9	RBT, SPL, SGR, SMB, CCF, BCR, CUT, TGM, RSF
HORSETOOTH RESERVOIR	W of Ft Collins	SP	LAR	1899.4	5430	A 8	RBT, SXW, MAC, SPL, LOC, WAL
HUGO WA PONDS	S of Hugo	AR	LIN	20.0	4740	E 12	RBT, BGL, CCF LMB, YPE
HUNTS LAKE	HUNTS CR, SW of Howard	AR	FRE	3.5	11320	F 7	CUT
I							
I-25 POND	SE of Ft Collins, N of exit 26	SP	LAR	10.0	4850	A 8	CCF, BGL, LMB
ICE LAKE	FALL R, NW of Alice	SP	CLE	12.0	12200	C 7	CUT
ICEBERG LAKE	McINTYRE CR, SW/Glendevey, Rawah	NP	LAR	5.7	11095	A 7	CUT
ICEBERG LAKE, NORTH	S BOULDER CR, SW of East Portal	SP	GIL	10.0	11640	C 7	CUT
IDAHO SPRINGS RES	SW/Idaho Spgs, aka CHICAGO CR R	SP	CLE	20.0	10617	C 7	CUT, BRK
IRON LAKE	IRON CR, ST LOUIS CR, FRASER R	CR	GRA	2.4	11440	C 7	CUT
ISABELLE LAKE	S ST VRAIN CR, W of Ward	SP	BOU	30.0	10868	B 7	CUT, RBT
ISLAND LAKE	W BRNCH LARAMIE, NW/Chambers L	NP	LAR	14.3	11128	A 7	CUT, BRK
ISLAND LAKE	BUCHANAN CR, E of Lk Granby	CR	GRA	18.6	11430	B 7	CUT, RBT
J							
JACKSON RESERVOIR	N of Wiggins, SP R #12	SP	MOR	2967.0	4438	B 10	RBT, WAL, SXW, BCR
JAMES PEAK LAKE	MAMMOTH CR, SW of Tolland	SP	GIL	10.0	11212	C 7	CUT
JEFFERSON LAKE	JEFFERSON CR, NW of Jefferson	SP	PAR	145.0	10687	D 7	RBT, MAC
JENNY LAKE	JENNY CR, NW of Tolland	SP	BOU	4.5	10917	C 7	RBT, BRK
JOE WRIGHT RESERVOIR	JOE WRIGHT CR, SW of Chambers Lake	SP	LAR	100+	9700	A 7	CUT
JOHN MARTIN RESERVOIR	E of Las Animas	AR	BEN	17500	3800	F 12	SXW, LMB, FLC, CCF, BCF, SBS
JOHN ROBERTSON PONDS (2)	POITREY CANYON, CHACUACO CR	AR	LAS	5.0	5700	H 11	CCF, LMB
JUMBO ANNEX	NE of Crook, SW of Jumbo Lk	SP	SED	200.0	3690	A 13	CCF, SMB, LMB, YPE, TGM
JUMBO RESERVOIR	NE of Crook	SP	SED	1570.0	3705	A 13	YPE, WAL, CCF, GSD, WAL, BCR, SXW
K							
KARVAL RESERVOIR	SE of Karvel	AR	LIN	24.0	5000	E 11	RBT, BGL, CCF, LMB, BCR, YPE
KELLY LAKE	KELLY CR, NE of Gould, State Forest	NP	JAC	21.0	10805	A 7	GOL
KING LAKE	S FK MID BOULDER, W of Eldora	SP	BOU	11.5	11431	C 7	CUT
KINNEY LAKE	S of Hugo	AR	LIN	7.0	5070	E 12	RBT, BGL, CCF, LMB
L							
LAGERMAN RESERVOIR	SW of Longmont	SP	BOU	115.0	5100	B 8	RBT
LAKE DOROTHEY	NE of Aston. SE of Trinidad	AR	LAS	10.0	7600	H 10	RBT
LAKE OF THE CLOUDS, LWR	SWIFT CR, W of Westcliffe	AR	CUS	10.5	11480	F 7	CUT
LAKE OF THE CLOUDS, MID	SWIFT CR, W of Westcliffe	AR	CUS	14.2	11560	F 7	CUT
LAKE OF THE CLOUDS, UPR	SWIFT CR, W of Westcliffe	AR	CUS	13.4	11640	F 7	CUT
LARAMIE LAKE	LARAMIE R, NE of Chambers Lk	NP	LAR	15.0	9327	A 7	RBT
LEFTHAND CR RESERVOIR	LEFT HAND CR, W of Ward	SP	BOU	100.0	10600	B 7	SPL, BRK, RBT
LILY LAKE, LOWER	HUERFANO R HDWTRS, N/Blanca Peak	AR	HUE	7.3	12350	G 8	CUT
LILY LAKE, UPPER	HUERFANO R HDWTRS, N/Blanca Peak	AR	HUE	4.0	12630	G 8	CUT
LINCOLN LAKE	BEAR CR #3, NE of Mt Evans	SP	CLE	12.8	11620	C 7	SPL
LITTLE BEAR LAKE	TOBIN CR, SW side Mt Blanca	RG	ALA	3.0	11500	G 7	CUT
LON HAGLER RESERVOIR	SW of Loveland	SP	LAR	200.0	5152	B 8	RBT, CCF, TGM, LMB
LONETREE RESERVOIR	NW of Berthoud	SP	LAR	502.0	5131	B 8	WAL, SXW, CCF, CRA
LONG DRAW RESERVOIR	LONG DRAW, SE of Chambers Lk	SP	LAR	242.3	10065	A 7	CUT
LONG LAKE	S ST VRAIN CR, W of Ward	SP	BOU	39.5	10521	B 7	CUT, RBT
LONG LAKE	HELL'S CANYON CR, E/Lk Granby	CR	GRA	5.0	9940	B 7	RBT, CUT
LONG LAKE	SW of La Veta	AR	HUE	2.0	7500	H 8	RBT
LONGS CANYON POND	SW side of TRINIDAD LK	AR	LAS	10.0	6230	H 9	CCF, LMB
LOST LAKE	BULL CR, NE of Red Feather Lks	SP	LAR	7.5	8040	A 7	RBT
LOST LAKE.	LARAMIE R, N of Chambers Lk	SP	LAR	27.4	9290	A 7	RBT
LOST LAKE	S FK MID BOULDER, SW of Hessie	SP	BOU	8.6	9786	C 7	RBT, BRK
LOST LAKE	HUERFANO R #3, N of Lily Lk	AR	HUE	3.2	12265	G 8	CUT
LOVELAND PASS LAKE	SW of Loveland Pass	CR	SUM	1.0	11835	C 7	RBT
M							
MACEY LAKE, LOWER	MACEY CR, SW of Westcliffe	AR	CUS	8.8	11506	F 7	CUT

Name	Location	Drainage	County	Size	Elev.	Map Coor.	Fish Likely to be Present
MACEY LAKE, SOUTH	MACEY CR, SW of Westcliffe	AR	CUS	16.9	11643	G 7	CUT
MACEY LAKE, WEST	MACEY CR, SW of Westcliffe	AR	CUS	12.5	11865	G 7	CUT
MANITOU RESERVOIR	TROUT CR, NW of Woodland Park	SP	TEL	16.0	7740	D 8	RBT, CUT
MARTIN LAKE	W of Walsenburg, Lathrop State Park	AR	HUE	206.0	6410	G 9	RBT, SXW, CCF, BCF, RSF
MARYS LAKE	FISH CR, SW of Lake Estes	SP	LAR	42.0	8046	B 7	RBT
MAYHEM POND #1	PURGATOIRE R	AR	LAS	2.0	5400	H 11	BGL, CCF, LMB
MAYHEM POND #2	PURGATOIRE R	AR	LAS	20	5400	H 11	BGL, CCF, LMB
MAYHEM POND #3	PURGATOIRE R	AR	LAS	2.0	5400	H 11	BGL
MEADOW CREEK RESERVOIR	10 miles E of Tabernash	CR	GRA	164.0	10000	B 7	RBT, BRK
MEDANO LAKE	MEDANO CR HDWTRS, Sand Dunes	RG	SAG	2.7	11680	F 8	CUT
MERIDITH RESERVOIR	SE of Ordway	AR	CRO	3220.0	4254	F 11	SXW, BCF, CCF, BGL, WAL, TGM
MICHIGAN LAKE, LOWER	MICHIGAN CR, NW of Jefferson	SP	PAR	3.7	11222	D 7	CUT
MICHIGAN LKS, UPPER	MICHIGAN R, S of Cameron Pass	NP	JAC	5.0	11240	B 7	CUT
MITCHELL LK, BIG	S ST VRAIN, NW of Ward	SP	BOU	13.8	10720	B 7	CUT, BRK
MITCHELL LK, LITTLE #1	S ST VRAIN, NW of Ward	SP	BOU	4.3	10720	B 7	CUT, BRK
MIZER POND	PINTADA CR, SE of Kim	AR	LAS	6.0	5200	H 12	BGL, LMB
MIZER POND, EAST	PINTADA CR, SE of Andrix	AR	BAC	6.0	5200	H 12	BGL, LMB
MONARCH LAKE	SE of Lk Granby	CR	GRA	160.0	8340	B 7	CUT, BRK
MONUMENT LAKE	MONUMENT CR, W of Monument	AR	ELP	40.0	6960	D 9	RBT, CUT, CCF, LMB
MONUMENT RESERVOIR	30 mi W of Trinidad	AR	LAS	100.0	8584	H 8	KOK, RBT, CUT, LOC
MORRAINE LAKE	S ST VRAIN CR, SW/Red Rock Lk	SP	BOU	2.0	10150	B 7	RBT, CUT
MOUNTAIN HOME RESERVOIR	TRINCHERA CR, SE of Ft Garland	RG	COS	631.0	8145	H 8	RBT, BRK, CUT
MURRAY LAKE	S CLEAR CR, SW of Georgetown	SP	CLE	11.4	12080	C 7	CUT
McCALLS LAKE	SE of Lyons on Hwy 66	SP	BOU	46.0	5179	B 8	CUT, RBT
McQUEARY LAKE	McQUEARY CR, WMS ER COLD R #4	CR	GRA	2.1	11040	C 7	CUT
N							
NEE GRONDE RES	NW of Lamar	AR	KIO	3490.0	3876	F 13	SAG, SXW, BCF, CCF, WAL
NEE NOSHE RES	NW of Lamar	AR	KIO	3696.0	3922	F 13	SXW, SAG, BCF, CCF, WAL
NEVA LAKE, LOWER	N FK MID BOULDER, S/Lk Dorothey	SP	BOU	10.0	11800	B 7	RBT
NICHOLS RESERVOIR	W MONUMENT CR, E/Ramparl Res	AR	ELP	20.0	8800	E 9	RBT, CUT
NORTH LAKE	30 mi W of Trinidad	AR	LAS	104.0	8583	H 8	RBT, SPL
NORTH STERLING RES	NW of Sterling, State Park	SP	LOG	2880.0	4050	A 12	CCF, SBS, WAL, SXW, SMB
O							
OLNEY SPRINGS PONDS	NE of Fowler	AR	CRO	6.0	4329	F 11	BGL, CCF, LMB
ORDWAY RESERVOIR	N of Ordway	AR	CRO	20.0	4340	F 11	RBT, BGL, SXW, LMB, CCF
OTERO POND	TIMPAS CR, W of La Juanta	AR	OTE	2.0	4200	G 11	BGL, CCF, LMB
P							
PAIUTE LAKE	THDRBLT CR, BUCHANAN CR	CR	GRA	12.0	10690	B 7	CUT
PALMER LAKE	MONUMENT CR, E City of Palmer Lake	AR	ELP	100+	7500	D 9	RBT, CUT
PARIKA LAKE	BAKER GULCH, NW of Grand Lake	CR	GRA	3.0	11450	B 7	CUT
PARVIN LAKE	S LONE PINE, Red Feather Group	SP	LAR	63.0	8200	A 7	RBT, LOC
PAWNEE LAKE	CASCADE CR, BUCHANAN CR	CR	GRA	11.0	10870	B 7	CUT, BRK
PETERSON LAKE	POUDRE R #5, N of Long Draw Res	SP	LAR	41.5	9492	A 7	CUT
PINEWOOD RESERVOIR	COTTONWOOD CR, NW of Carter Lk	SP	LAR	97.0	6577	B 8	RBT
PIONEER LAKE	PIONEER CR, W side Mt Blanca	RG	ALA	6.9	12000	G 7	CUT
PLATTE CANYON RES	SP River, at Kassler	SP	DOU	59.0	5519	D 8	RBT
PONY LAKE	Peaceful Valley	SP	BOU	3.6	9650	B 8	RBT
POUDRE UNIT LAKE	POUDRE R #4, Poudre River SFU	SP	LAR	4.0	7720	A 7	RBT
PREWITT RESERVOIR	S of Merino	SP	WAS	2431.0	4100	B 12	CCF, WAL, SXW
PRONGER POND	Tamarack SWA, S of Crook	SP	LOG	4.0	3600	A 8	BGL, LMB
PUEBLO RESERVOIR	W of Pueblo	AR	PUE	3000.0	4880	F 9	RBT, WAL, SXW, FLC, BCF, CCF, SMB, LMB
PURSLEY POND	TWO BUTTES CR, NE of Kim	AR	LAS	8.0	4100	H 12	BGL, CCF
Q							
QUINCY RESERVOIR	3 mi E of Cherry Creek Reservoir	SP	ARA	161.0	5699	C 9	RBT, TGM
R							
RAINBOW LAKE	N LAKE CR, W of Hillside	AR	CUS	10.7	10400	F 7	CUT
RAINBOW LAKE #2	CARIBOU CR, S of Silver Lk	SP	BOU	2.9	10180	B 7	RBT
RAMAH RESERVOIR	NE of Calhan	AR	ELP	170.0	6100	D 10	BGL, CCF, NPK, WAL, BGL
RAMPART RESERVOIR	W MONUMENT CR, E Woodland Park	AR	ELP	500.0	9000	E 9	MAC, RBT, CUT
RANGER LAKE. LOWER	MICHIGAN R, SE of Gould	NP	JAC	4.3	9233	A 7	RBT
RANGER LAKE, UPPER	MICHIGAN R, SE of Gould	NP	JAC	8.5	9243	A 7	RBT
RAWAH LAKE #3	RAWAH CR, NW of Chambers Lk	NP	LAR	22.0	10873	A 7	BRK
RAWAH LAKE #4	RAWAH CR, NW of Chambers Lk	NP	LAR	26.0	11474	A 7	CUT
RED DEER LAKE	MID ST VRAIN, NW of Ward	SP	BOU	16.0	10372	B 7	LOC, RBT, CUT, BRK
RED ROCK LAKE	S ST VRAIN CR, W of Ward	SP	BOU	6.5	10160	B 7	RBT
RITO ALTO LAKE	RITO ALTO CR, SAN LUIS CR	RG	SAG	4.0	11240	F 7	CUT
ROCK HOLE LAKE	Rawah Wild Area	NP	LAR	6.0	10800	A 7	CUT
ROGERS LAKE	Rawah Wild Area	NP	LAR	4.3	11320	A 7	CUT
ROGERS PASS LAKE	S BOULDER CR, SW of East Portal	SP	GIL	5.6	11120	C 7	CUT
ROLFS LAKE #1	Elephant Mountain	SP	LAR	15.7	9645	A 7	CUT
ROLFS LAKE #2	Elephant Mountain	SP	LAR	11.0	9600	A 7	CUT
ROOSEVELT LAKE, LOWER	TRUESDELL CR, SE of Mt Evans	SP	CLE	1.5	11742	C 7	CUT
ROOSEVELT LAKE. UPPER	TRUESDELL CR, SE of Mt Evans	SP	CLE	1.7	11742	C 7	CUT
ROSEMONT RESERVOIR	E BEAVER CR, E of Cripple Creek	AR	TEL	90.0	9640	E 9	CUT, SPL
ROUND LAKE	HELLS CANYON CR, E/Lk Granby	CR	GRA	2.5	11510	B 7	RBT
RUBY JEWEL LAKE	S FK CANADIAN R, NE of Gould	NP	JAC	4.0	11240	A 7	CUT
RYAN POND (POTTS)	Rocky Ford Bird Farm	AR	OTE	2.0	4100	F 11	BGL, CCF, LMB, BCR
RYAN POND, EAST	Rocky Ford Bird Farm	AR	OTE	5.0	4100	F 11	BGL, CCF, LMB
RYAN POND, MIDDLE	Rocky Ford Bird Farm	AR	OTE	2.0	4100	F 11	BGL, CCF, LMB
RYAN POND, NORTH	Rocky Ford Bird Farm	AR	OTE	1.0	4100	F 11	BGL, CCF
RYAN POND, SE	Rocky Ford Bird Farm	AR	OTE	5.0	4100	F 11	BGL, CCF, LMB
RYAN POND, WEST	Rocky Ford Bird Farm	AR	OTE	2.0	4100	F 11	BGL, CCF

Name	Location	Drainage	County	Size	Elev.	Map Coor.	Fish Likely to be Present
S							
S BRANCH BRUSH CR LAKE	N BRUSH CR, NW of Westcliffe	AR	CUS	2.0	11540	F7	CUT
SAN ISABEL LAKE	ST CHARLES R #3, NW of Rye	AR	CUS	30.0	8474	G 8	RBT, SPL, CUT
SAN ISABEL LAKE	SAN ISABEL CR, SE of Rito Alto Lk	RG	SAG	5.9	11600	F 7	CUT
SAN LUIS LAKE	SAN LUIS CR, NE of Alamosa, State P	RG	COS	4571.0	8317	H 8	WAL, NPK, YPE
SAND CR BEAVER PONDS	NE of Sand Cr Pass, into Wyo	SP	LAR	2.0	9040	A 7	LOC, BRK
SAND CR LAKE, LITTLE	LIT SAND CR	RG	SAG	12.6	11088	G 7	CUT, BRK
SAND CR LAKE, LOWER	SAND CR	RG	SAG	62.8	11471	G 7	CUT, BRK
SAND CR LAKE, UPPER	SAND CR	RG	SAG	42.8	11745	G 7	CUT, BRK
SANTA MARIA POND	N FK SP #IB	SP	PAR	1.0	8550	D 7	RBT
SAWHILL POND #1	BOULDER CR, N of Valmont Res	SP	BOU	11.7	5140	B 8	WAL, SXW, CCF
SAWHILL POND #1A	BOULDER CR, N of Valmont Res	SP	BOU	1.0	5140	B 8	CCF
SAWHILL POND #2	BOULDER CR, N of Valmont Res	SP	BOU	9.4	5140	B 8	WAL, SXW
SHADOW MOUNTAIN RES	COLO R #11,S of Grand Lake	CR	GRA	1356.0	8367	B 7	RBT, GRA, LOC, BRK, CUT, MAC
SHELF LAKE	SMELTER GULCH, W of Mt Evans	SP	CLE	9.0	11970	C 7	CUT
SHERWIN LAKE	FALL R, W of Alice	SP	CLE	8.6	11087	C 7	RBT, SPL, BRK, LOC
SILVER DOLLAR LAKE	S CLEAR CR, W of Mt Evans	SP	CLE	18.6	11950	C 7	CUT
SILVER LAKE	N LAKE CR, SW of Hillside	AR	CUS	5.2	11950	F 7	CUT
SKAGWAY RESERVOIR	W BEAVER CR, SE of Victor	AR	TEL	114.0	8880	E 8	RBT, BRK, MAC
SKYSCRAPER LAKE	JASPER CR, W of Eldora	SP	BOU	12.9	11221	C 7	CUT, RBT
SLATER LAKE	FALL R. W of Alice	SP	CLE	7.2	11385	C 7	CUT, BRK
SLIP LAKE	Poudre Pass	SP	LAR	3.0	10920	B 7	RBT, CUT
SMITH RESERVOIR	TRINCHERA CR, S of Blanca	RG	COS	700.0	7720	H 7	RBT, CCF
SNOW LAKE	MICHIGAN R, S of Cameron Pass	NP	JAC	17.2	11516	B 7	RBT, CUT
SPINNEY MTN RES	S PLATTE R, W/Lake George	SP	PAR	2520.0	8676	E 7	RBT, CUT, LOC, KOK, NPK
SQUARE TOP LAKE, LOWER	DUCK CR, W of Mt Evans	SP	CLE	8.6	12046	C 7	CUT
SQUARE TOP LAKE, UPPER	DUCK CR, W of Mt Evans	SP	CLE	10.0	12240	C 7	CUT
ST ANNS POND	N FK S Platte	SP	PAR	1.0	8560	D 7	RBT
ST LOUIS LAKE	ST LOUIS CR, FRASER R	CR	GRA	3.2	11531	C 7	CUT
STABILIZATION RESERVOIR	COSTILLA R, SW of San Luis	RG	COS	43.0	7816	H 7	RBT, SMB
STALKER LAKE	CHIEF CR, W of Wray	RE	YUM	26.6	3566	B 14	RBT, CCF, BGL, LMB
STANLEY RESERVOIR	STANLEY CANYON. W/USAFA	AR	ELP	12.0	8800	E 9	RBT, CUT
STONE LAKE	HELLS CANYON CR, E/Lk Granby	CR	GRA	6.0	10643	B 7	RBT
STORM LAKE	JASPER CR, NW of Eldora	SP	BOU	7.0	11440	C 7	CUT
STOUT CR LAKE #3	STOUT CR, SW of Howard	AR	FRE	2.0	11850	F 7	CUT
STOUT CR LAKE, LOWER	STOUT CR, SW of Howard	AR	FRE	19.5	11720	F 7	CUT
STOUT CR LAKE, UPPER	STOUT CR, SW of Howard	AR	FRE	12.3	11800	F 7	CUT
STRONTIA SPRINGS RES	S PLATTE R, SW of Chatfield Res	SP	JEF	96.0	5990	D 8	RBT, KOK, SPL
SUMMIT LAKE	BEAR CR #3, N side of Mt Evans	SP	CLE	32.8	12830	C 7	RBT
T							
TAMARACK RANCH POND	S of Crook	SP	LOG	2.0	3680	A 13	CCF, BGL, LMB, RBT
TARRYALL RES	TARRYALL CR, E/Fairplay 20 mi	SP	PAR	175.0	8860	D 7	RBT, LOC, KOK
TEACUP LAKE	COTTON CR, NE of Cotton Lk	RG	SAG	3.0	11700	E 7	CUT
THUNDERBOLT LAKE	BUCHANAN CR, E of Lk Granby	CR	GRA	4.2	11045	B 7	CUT
THURSTON RESERVOIR	N of Lamar	AR	PRO	173.0	3797	F 13	SXW, CCF, LMB
TIMBER LAKE	W BRCH LARAMIE, NW/Chambers Lk	NP	LAR	13.0	11020	A 7	RBT, CUT
TRAP LAKE	TRAP CR, SE of Chambers Lk	SP	LAR	13.0	9975	A 7	RBT
TRAPPERS LAKE	CABIN CR, N FK WHITE R	WR	GAR	???	9926	A 7	CUT
TRAPPERS LAKE, LITTLE	CABIN CR, N FK WHITE R	WR	GAR	20.0	9926	A 7	CUT
TRINIDAD RESERVOIR	PURGATOIRE R, SW of Trinidad	AR	LAS	1045.0	6225	H 9	RBT, SXW, CUT, SMB, BCF, BCR, LMB, CCF
TURKS POND	E of Springfield	AR	BAC	25.0	3970	G 13	BGL, CCF, BCR, RSF, LMB
TWIN CRATER LAKE, LOWER	W BRCH LARAMIE, NW/Chambers Lk	NP	LAR	7.2	11043	A 7	CUT
TWIN CRATER LAKE, UPPER	W BRCH LARAMIE, NW/Chambers Lk	NP	LAR	17.2	11047	A 7	CUT
TWIN LAKE, EAST	McINTYRE CR, SW of Glendevey	NP	LAR	20.5	9453	A 7	BRK, RBT
TWIN LAKE, UPPER	McINTYRE CR, SW of Glendevey	NP	LAR	8.6	10610	A 7	RBT, BRK
TWO BUTTES PONDS	NE of Springfield	AR	BAC	6.0	4000	G 13	CCF, BGL, LMB
U							
UNION RESERVOIR	E of Longmont, also CALKINS RES	SP	WEL	743.6	4956	B 8	RBT, SXW, CCF, BCR
UPPER LAKE	HELLS CANYON CR, E/Lk Granby	CR	GRA	7.0	10730	B 7	CUT
URAD MINE LAKE	WOODS CR, W of Empire	SP	CLE	31.0	10644	C 7	RBT, CUT, BRK
UTE LAKE, LITTLE, LOWER	Blanca Trinchera Ranch	RG	COS	3.0	11600	G 8	CUT
UTE LAKE, LITTLE, UPPER	Blanca Trinchera Ranch	RG	COS	2.0	11800	G 8	CUT
V							
VENABLE LK, LOWER	VENABLE CR, SW of Westcliffe	AR	CUS	8.5	11985	F 7	CUT
VENABLE LK, UPPER	VENABLE CR, SW of Westcltfe	AR	CUS	5.0	12070	F 7	CUT
W							
WAHATOYA RESERVOIR	CUCHARAS CR, E of La Veta	AR	HUE	6.5	7110	G 9	CUT, SPL, LOC, RBT
WALDEN POND, SOUTH	SOUTH POND, N of Valmont Res	SP	BOU	6.3	5110	B 8	LMB
WATANGA LAKE	ROARING FORK CR, E/Lk Granby	CR	GRA	2.8	10790	B 7	CUT
WATSON LAKE	POUDRE R, NE of Bellvue	SP	LAR	40.0	5160	A 8	RBT, LOC, BRK
WELLINGTON RES #4	NW of Wellington	SP	LAR	103.0	5228	A 8	RBT, CCF, LMB, YPE, WAL
WEST CREEK LAKE	WEST CR, SW of Howard	AR	FRE	8.3	11675	F 7	CUT
WEST LAKE	S LONE PINE, Red Feather Group	SP	LAR	24.9	8246	A 7	RBT, LOC
WILDERNESS ON WHEELS POND	KENOSHA CR, base/Kenosha Pass	SP	PAR	1.0	9200	D 7	RBT
WILLOW CREEK LAKE, LOWER	WILLOW CR, SAN LUIS CR	RG	SAG	19.8	11564	G 7	CUT
WILLOW CREEK RESERVOIR	WILLOW CR, N of Granby	CR	GRA	303.0	8130	B 7	RBT, BRK
WOLF LAKE	CUCHARAS CR, above Bear Lake	AR	HUE	0.5	11100	H 8	RBT
WOODLAND LAKE	unnmd to JASPER CR, W/Eldora	SP	BOU	10.5	10972	C 7	GRA, RBT
Y							
YANKEE DOODLE LAKE	JENNY CR, NW of Tolland	SP	BOU	5.7	10711	C 7	RBT

Appendix 2

Stocked
Lakes and Reservoirs
of
Western Colorado

Note: For easy reference the state has been divided into two parts, eastern and western sections. A reference point (106°) for the eastern and western sections would be at Lake Dillon at I-70 in Summit County. An imaginary line (106°) running north and south from this point divides the listed lakes into east and west sections.

DOW Stocked Lakes Code Explanation

Major River Drainage Code

AR	ARKANSAS RIVER
CR	COLORADO RIVER
GR	GREEN RIVER
GU	GUNNISON RIVER
NP	NORTH PLATTE RIVER
RE	REPUBLICAN RIVER
RG	RIO GRANDE RIVER
SJ	SAN JUAN RIVER
SP	SOUTH PLATTE RIVER
WR	WHITE RIVER
YP	YAMPA RIVER

County Codes

ADA	ADAMS
ALA	ALAMOSA
ARA	ARAPAHOE
ARC	ARCHULETA
BAC	BACA
BEN	BENT
BOU	BOULDER
CHA	CHAFFEE
CLE	CHEYENNE
CON	CONEJOS
COS	COSTILLA
CRO	CROWLEY
CUS	CUSTER
DEL	DELTA
DEN	DENVER
DOL	DOLORES
DOU	DOUGLAS
EAG	EAGLE
ELB	ELBERT
ELP	EL PASO
FRE	FREMONT
GAR	GARFIELD
GIL	GILPIN
GRA	GRAND
GUN	GUNNISON
HIN	HINSDALE
HUE	HUERFANO
JAC	JACKSON
JEF	JEFFERSON
KIO	KIOWA
KIT	KIT CARSON
LAK	LAKE
LAP	LA PLATA
LAR	LARIMER
LAS	LAS ANIMAS
LIN	LINCOLN
LOG	LOGAN
MES	MESA
MIN	MINERAL
MOR	MOFFAT
MTZ	MONTEZUMA
MON	MONTROSE
MOR	MORGAN
OTE	OTERO
OUR	OURAY
PAR	PARK
PHI	PHILLIPS
PIT	PITKIN
PRO	PROWERS
PUE	PUEBLO
RBL	RIO BLANCO
RGR	RIO GRANDE
ROU	ROUTT
SAG	SAGUACHE
SNJ	SAN JUAN
SNM	SAN MIGUEL
SED	SEDGWICK
SUM	SUMMIT
TEL	TELLER
WAS	WASHINGTON
WEL	WELD
YUM	YUMA

Fish Species Codes

BCF	CATFISH — BLUE
BCR	BLACK CRAPPIE
BGL	BLUEGILL
BHD	BULLHEAD
BRK	BROOK TROUT
CCF	CATFISH — CHANNEL
CRA	CRAPPIE
CUT	CUTTHROAT
FLC	CATFISH — FLATHEAD
GOL	GOLDEN TROUT
GRA	GRAYLING
KOK	KOKANEE SALMON
LMB	BASS — LARGE MOUTH
LOC	BROWN TROUT
LXB	TIGER TROUT
MAC	MACKINAW (LAKE) TROUT
NPK	NORTHERN PIKE
RBT	RAINBOW TROUT
SAG	SAUGEYE
SBS	STRIPED BASS
SGR	SAUGER
SMB	BASS — SMALL MOUTH
SPD	BASS — SPOTTED
SPL	SPLAKE
SQF	SQUAWFISH
SXW	WIPER
TGM	TIGER MUSKIE
WAL	WALLEYE
WBA	WHITE BASS
WHF	WHITE FISH
YPE	YELLOW PERCH

Heading Explanation

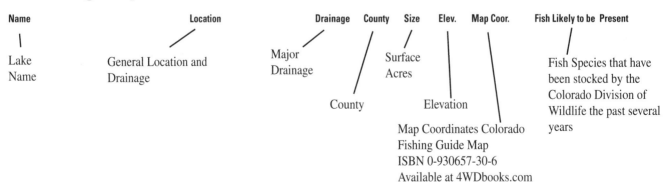

Name	Location	Drainage	County	Size	Elev.	Map Coor.	Fish Likely to be Present
Lake Name	General Location and Drainage	Major Drainage	County	Surface Acres	Elevation	Map Coordinates Colorado Fishing Guide Map ISBN 0-930657-30-6 Available at 4WDbooks.com	Fish Species that have been stocked by the Colorado Division of Wildlife the past several years

Name	Location	Drainage	County	Size	Elev.	Map Coor.	Fish Likely to be Present
A							
ADAMS LAKE	CANYON CR, SW of Heart Lake	CR	GAR	31.0	10796	C 4	CUT
AGUA FRIA LAKE	BEAVER CR, S of Rainbow Lakes	NP	JAC	27.6	10040	A 5	MAC, BRK
ALAN LAKE	N FK CLEAR CR, NW of Winfield	AR	CHA	2.3	12100	E 5	CUT
ALBERT LAKE	CHEDSEY CR, N/Buffalo Pass	NP	JAC	10.7	10177	A 5	CUT, RBT
ALBERTA PARK RESERVOIR	TRIB TO PASS CR, S FK RG	RG	MIN	40.0	10203	H 5	BRK, RBT
ALDER LAKE	NINEGAR CR, SE/Seymour Lk SWA	NP	JAC	3.0	9500	B 6	RBT, CUT
ALDRICH LAKE #1	MARTIN CR, MILK CR, NE/Meeker	YP	RBL	50.7	7440	B 3	CUT
ALDRICH LAKE #2	MARTIN CR, MILK CR, NE/Meeker	YP	RBL	13.5	7505	B 3	CUT
ALDRICH LAKE #3	MARTIN CR, MILK CR, NE/Meeker	CR	RBL	5.0	7555	B 3	CUT
ALEXANDER LAKE	WARD CR, E of Island Lk	GU	DEL	40.8	10130	D 3	RBT
ALLEN BASIN RESERVOIR	HUNT CR, W of Yampa	YP	ROU	60.0	8680	B 4	RBT, BRK
ALTA LAKE, LOWER	HOWARDS FK SNM R, SW/Telluride	DO	SNM	9.9	11216	G 3	BRK, RBT
ALVER JONES LAKE	ROUGH CR, CONEJOS R #2	RG	CON	16.1	11180	H 5	CUT
AMERICAN LAKE	DEVANEY CR, CASTLE CR	CR	PIT	2.8	11365	D 5	RBT
ANDERSON LAKE	LINCOLN CR, ROARING FK R	CR	PIT	8.6	11840	D 5	RBT
ANDERSON RESERVOIR	BEAR CR, N FK WHITE, NW/Trapper	WR	GAR	7.2	9340	B 4	BRK, RBT
ANDREWS LAKE	Molas Pass, SW of Silverton	SJ	SNJ	8.0	10745	G 3	RBT, BRK
ANGLEMEYER LAKE	N COTTONWOOD CR, N/Kroenke Lk	AR	CHA	1.5	11338	E 6	CUT
ANN LAKE	MID FK CONEJOS R, SW of Platoro	RG	CON	15.6	11910	H 5	CUT
ANN LAKE	S FK CLEAR CR, SW of Winfield	AR	CHA	18.6	11800	B 6	CUT
ANNIE LAKE	ROCK CR, VALLECITO CR #2	SJ	LAP	3.3	12200	G 4	CUT
ARCHULETA LAKE	S FK RG R, W of Big Meadows Res	RG	MIN	4.0	11720	G 5	CUT, BRK
ARTHUR LAKE	N FK S ARKANSAS R	AR	CHA	5.0	12450	E 6	CUT
ASPEN LAKE	CIMARRON R, NE/Silverjack Res	GU	GUN	3.0	8820	F 3	RBT
ATKINSON RESERVOIR	BIG CR, PLATEAU CR, Grand Mesa	CR	MES	87.4	10120	D 3	RBT
AVALANCHE LAKE	AVALANCHE CR, CRYSTAL R	CR	PIT	8.5	10695	D 4	CUT
AVERY LAKE	WHITE R #2, W of Buford	WR	RBL	245.0	6985	C 3	RBT, CUT
AXIAL BASIN LAKE	Hwy 13, NE of Axial	YP	MOF	25.0	6260	B 3	RBT
B							
BALDWIN LAKE, UPPER	BALDWIN CR, CHALK CR, S/Alpine	AR	CHA	21.0	12100	E 6	CUT
BALDY LAKE	BALDY BR, LONG BR, TOMICHI CR	GU	SAG	5.2	11266	F 6	RBT
BARON RESERVOIR	WARD CR, E of Alexander Lk	GU	DEL	98.3	10125	D 3	RBT
BATTLEMENT RES, BIG	DIRTY GEORGE CR, SW Island Lk	GU	DEL	41.0	10080	D 2	RBT, BRK
BATTLEMENT RES., LIT.	DIRTY GEORGE CR, SW Island Lk	GU	DEL	15.6	10050	D 2	RBT, BRK
BATTLEMENT RESERVOIR #1	BATTLEMENT CR, S of Rulison	CR	GAR	11.9	10180	D 3	CUT
BATTLEMENT RESERVOIR #2	BATTLEMENT CR, S of Rulison	CR	GAR	4.8	10190	D 3	CUT
BATTLEMENT RESERVOIR #3	BATTLEMENT CR, S of Rulison	CR	GAR	4.1	10190	D 3	CUT
BATTLEMENT RESERVOIR #4	BATTLEMENT CR, S of Rulison	CR	GAR	6.0	10150	D 3	CUT
BATTLEMENT RESERVOIR #5	BATTLEMENT CR, S of Rulison	CR	GAR	2.0	10190	D 3	CUT
BEAR LAKE	aka YAMPA RES, SW of Yampa	YP	GAR	46.7	9720	B 4	RBT, BRK
BEAR LAKE	BEAR CR, N of Turquoise Lk	AR	LAK	12.3	11050	D 6	RBT
BEAR LAKE	HORN FK CR, N COTTONWOOD CR	AR	CHA	18.5	12377	E 6	CUT
BEAR LAKE	SADDLE CR, CONEJOS R. S/Platoro	RG	CON	18.3	11520	H 5	CUT, RBT
BEAR LAKE, LITTLE	BEAR CR, N of Turquoise Lk	AR	LAK	2.3	11075	D 6	CUT
BEAR LAKE, NORTH	BEAR CR, W of Lake John	NP	JAC	10.1	10330	A 5	CUT, LOC
BEAR LAKE, UPPER	BEAR CR, W of Lake John	NP	JAC	5.7	10984	A 5	CUT
BEAVER CREEK RESERVOIR	BEAVER CR, S of South Fork	RG	RGR	113.9	8850	G 5	KOK, RBT, LOC
BEAVER DAM LAKE	SHEEPHORN CR, E of Slate Brldge	CR	EAG	2.0	9560	C 6	CUT
BEAVER FLATTOPS PONDS	S FK WMS FK Y, SE/Indian Run SW	YP	ROU	40.0	9320	B 4	BRK
BEAVER LAKE	BEAVER CR, S of Avon	CR	EAG	7.0	9746	C 5	BRK, CUT
BEAVER LAKE	CIMARRON R, N/Silverjack Res	GU	GUN	9.9	8740	F 3	RBT
BEAVER LAKE	CRYSTAL R, E of Marble	CR	GUN	25.0	7956	D 4	RBT, BRK
BEAVER LAKE	Mesa Lake Resort, Grand Mesa	CR	MES	6.0	9820	D 2	RBT, BRK, LOC
BEAVER LAKE	TRIB TO ELK CR, CONEJOS R #2	RG	CON	4.2	9560	H 6	CUT
BELAIRE PONDS	old BELAIRE SFU, SW of Buford	WR	RBL	1.0	6960	C 3	CUT
BENCH LAKE #1	HOMESTAKE CR, EAGLE R	CR	PIT	3.0	11140	D 6	CUT
BENCH LAKE #2	HOMESTAKE CR, EAGLE R	CR	PIT	12.0	11220	D 6	CUT
BERRY LAKE	BERRY CR, S FK WMS FK YAMPA R	YP	RBL	5.0	8280	B 4	RBT
BETTY LAKE	ROCK CR, VALLECITO CR #2	SJ	LAP	8.0	12210	G 4	CUT
BETTY LAKE	SPRING CR, TAYLOR R #1	GU	GUN	2.3	11870	E 5	CUT
BIG CREEK LAKE	BIG CR, ELK R, NE of Mad Creek	YP	ROU	8.2	10620	A 5	CUT
BIG CREEK LAKE, LOW	S FK BIG CR, SW of Pearl	NP	JAC	350.7	8997	A 5	LOC, RBT, TGM, MAC
BIG CREEK LAKE, UPPER	S FK BIG CR, SW of Pearl	NP	JAC	101.2	9009	A 5	RBT
BIG CREEK RESERVOIR	BIG CR, GILL CR, WEST CR	DO	MES	35.0	8240	E 1	CUT, RBT, LOC
BIG FISH LAKE	BIG FISH CR, N FK WHITE R	WR	GAR	20.0	9388	C 4	CUT, RBT
BIG HORN LAKE	LONE PINE CR, W of Lake John	NP	JAC	13.8	10106	A 5	CUT
BIG HORN LAKE, UPPER	LONE PINE CR, W of Lake John	NP	JAC	4.0	10200	A 5	CUT
BIG LAKE	LK FK CONEJOS R, S of Platoro	RG	CON	11.9	9840	H 5	CUT
BIG MEADOWS RESERVOIR	Grand Mesa	CR	MES	40.0	9800	D 3	RBT, BRK
BIG MEADOWS RESERVOIR	S FK RG, SW of South Fork	RG	MIN	113.9	9200	G 5	RBT, BRK
BILLINGS LAKE	N FK S ARKANSAS R, NW/Garfield	AR	CHA	6.0	11700	E 6	CUT
BLACK LAKE #1, UPPER	BLACK GORE CR, N of Vail Pass	CR	EAG	15.0	10470	C 6	RBT
BLACK LAKE #2, LOWER	BLACK GORE CR, N of Vail Pass	CR	EAG	12.5	10400	C 6	RBT
BLACK MOUNTAIN LAKE	N CLEAR CR, SE/Continental Res	RG	HIN	5.9	11200	G 4	CUT

Western Colorado Lakes

Name	Location	Drainage	County	Size	Elev.	Map Coor.	Fish Likely to be Present
BLACK PINE RESERVOIR	LOBE CR, WEST CR, DOLORES R	DO	MES	7.5	9200	E 1	RBT
BLAIR LAKE	PATTERSON CR, S FK WHITE R	WR	GAR	55.0	10466	C 4	CUT, RBT, BRK
BLODGETT LAKE	CROSS CR HDWTRS, SW/Minturn	CR	EAG	25.0	11665	D 5	CUT
BLUE LAKE	EL RITO AZUL CR, upper CONEJOS	RG	CON	49.4	11463	H 5	CUT, BRK
BLUE LAKE	KANNAH CR, NW of Chambers Res	GU	MES	8.0	8990	E 2	CUT
BLUE LAKE	KEYSER CR, NW of Glenwood Spgs	CR	GAR	5.3	10430	C 4	BRK
BLUE LAKE	N FK LAKE CR, NW of Twin Lakes	AR	LAK	23.4	12495	D 5	CUT
BLUE LAKE	N FK S ARKANSAS RIVER	AR	CHA	0.8	12400	E 6	RBT
BLUE LAKE	OH-BE-JOYFUL CR, SLATE R	GU	GUN	3.0	11056	E 4	CUT, RBT
BLUE LAKE, LOWER	E DALLAS CR, W of Ouray	GU	OUR	12.4	10980	F 3	CUT, RBT
BLUE LAKE, MIDDLE	E DALLAS CR, W of Ouray	GU	OUR	5.0	11500	F 3	CUT, RBT
BLUE LAKE, UPPER	E DALLAS CR, W of Ouray	GU	OUR	9.3	11740	F 3	CUT
BLUE MESA RESERVOIR	GUNNISON R #4, W of Gunnison	GU	GUN	9000.0	7528	F 4	KOK, RBT, LOC, MAC
BOLAM PASS LAKE	HERMOSA CR HDWTRS, ANIMAS R	SJ	SNJ	2.8	11093	G 3	BRK
BONHAM RESERVOIR	BIG CR, PLATEAU CR, Grand Mesa	CR	MES	87.5	9790	G 3	CUT
BONITA RESERVOIR'	E FK MARCOTT, N/Cedar Mesa Res	GU	DEL	23.0	10020	D 3	CUT
BOOTH LAKE	BOOTH CR, GORE CR, NE of Vail	CR	EAG	5.0	11470	D 3	CUT
BORAH LAKE	W BRUSH CR, W of Sylvan Lk	CR	EAG	1.5	10872	C 5	RBT, BRK
BOULDER LAKE #2	BOULDER CR, NW of Dillon	CR	SUM	4.9	11200	E 5	CUT, BRK
BOULDER LAKE #3	BOULDER CR, NW of Dillon	CR	SUM	14.0	10984	C 6	CUT, BRK
BOULDER LAKE #4	BOULDER CR, NW of Dillon	CR	SUM	5.2	11340	C 6	CUT, BRK
BOULDER LAKE #5	BOULDER CR, NW of Dillon	CR	SUM	7.7	11560	C 6	CUT, BRK
BOULDER LAKE #6	BOULDER CR, NW of Dillon	CR	SUM	1.4	11900	C 6	CUT, BRK
BOULDER LAKE	BIG FISH CR, N FK WHITE R	WR	GAR	4.0	9790	A 3	RBT
BOULDER LAKE	GOLD CR. QUARTZ CR. NW/Pitkin	GU	GUN	8.9	11580	C 4	BRK
BOWL OF TEARS LAKE	E CROSS CR, SW of Red Cliff	CR	EAG	20.0	12001	D 6	CUT
BRADY LAKE #1	SOPRIS CR, HOMESTAKE CR	CR	EAG	1.0	10850	D 6	CUT
BRADY LAKE #2	SOPRIS CR, MISSOURI CR	CR	EAG	8.9	10984	D 6	CUT
BROOKLYN GULCH LK., UPPER	NEW YORK CR, SE of Aspen	CR	PIT	2.9	12280	D 6	CUT
BROWN LAKE, LOWER	S CLEAR CR, W of Creede	RG	HIN	70.0	9800	G 4	RBT, BRK
BROWN LAKE, UPPER	S CLEAR CR, W of Creede	RG	HIN	109.9	9803	G 4	RBT, BRK
BRUSH CREEK LAKE #1	BRUSH CR, S of Green Mtn Res	CR	SUM	8.5	11560	C 6	RBT
BUBBLE LAKE	BLACK CR, Green Mtn Res	CR	SUM	14.3	11300	C 6	CUT
BUCK LAKE	S FK DERBY CR, W of Burns	CR	GAR	11.2	11040	C 4	CUT
BUCKEYE LAKE	BUCKEYE GUL, NE of Leadville	AR	LAK	1.7	11900	D 6	CUT
BUCKEYE RESERVOIR	BUCKEYE CR, NW of Paradox	DO	MON	85.0	7700	F 1	RBT
BUCKHORN LAKE, BIG	BEATON CR, E of Colona	GU	MON	16.3	9720	F 3	RBT, BRK
BUCKHORN LAKE, LITTLE	BEATON CR, E of Colona	GU	MON	12.6	9745	F 3	RBT, BRK
BUCKLES LAKE	BIG BRANCH CR, NE of Chromo	SJ	ARC	29.6	9518	H 5	RBT, BRK, CUT
BUFFALO LAKE	CROSS CR, EAGLE R, SW/Minturn	CR	EAG	2.0	10910	D 5	CUT
BULL CREEK RES #1	BULL CR, PLATEAU CR, Grand Mesa	CR	MES	10.0	10140	D 2	RBT
BULL CREEK RES #2	BULL CR, PLATEAU CR, Grand Mesa	CR	MES	60.0	10140	D 2	RBT
BULL CREEK RES #5	BULL CR, PLATEAU CR, Grand Mesa	CR	MES	18.0	9640	D 2	RBT, CUT
BULL LAKE	S FK DERBY CR, W of Burns	CR	GAR	3.7	11030	C 4	CUT
BULLION KING LAKE	PORPHYRY GUL, MINERAL CR	SJ	SNJ	3.0	12570	G 3	RBT
BUNDY LAKE	NINEGAR CR, S of Two Ledge Res	NP	JAC	3.0	9840	B 6	RBT, BRK
BUTTS LAKE	WARD CR, N of Big Eggleston Res	GU	DEL	22.7	10500	D 3	CUT
C							
CAMEL LAKE	SAND CR, TROUT CR	YP	RBL	4.0	10080	B 4	BRK
CAMP HALE POND	EAGLE R #4, SE of Red Cliff	CR	EAG	9.5	9270	D 6	RBT
CAPITOL LAKE	CAPITOL CR, SW/Snowmass Village	CR	PIT	21.7	11590	D 5	CUT
CARBONATE LAKE	BALDWIN GUL, CHALK CR. SE/Alpine	AR	CHA	3.2	12040	E 6	CUT
CARP LAKE	Grand Mesa, E of Island Lk	GU	DEL	11.9	10260	D 3	RBT
CARSON LAKE	Grand Mesa	CR	MES	35.0	10000	D 3	BRK
CASEY POND	BURGESS CR, SE/Steamboat Spgs	YP	ROU	3.0	6800	A 5	RBT
CASTILLEJA LAKE	FLORIDA R HDWTRS, NE/City Res	SJ	LAP	9.8	12200	G 3	CUT
CASTILLEJA LAKE, EAST	HDWTRS FLORIDA R, NE/City Res	SJ	LAP	5.9	12110	G 3	CUT
CATARACT LAKE	CATARACT GUL, S of Sherman	GU	HIN	17.0	12082	G 4	BRK
CATARACT LAKES	CATARACT CR, SW/Green Mtn Res	CR	SUM	12.5	10590	C 6	CUT, BRK, LOC, RBT
CATHEDRAL LAKE	PINE CR, ROARING FK R, S/Aspen	CR	PIT	34.0	11866	D 5	CUT, RBT, BRK
CAUSEWAY LAKE	E FK WMS FK Y, SW of Yampa	YP	GAR	4.8	10420	B 4	CUT
CAUSEWAY LAKE, LITTLE	BEAR R, NW of Stillwater Res	YP	GAR	15.0	10740	B 4	CUT, BRK
CEDAR MESA RESERVOIR	BONITA CR, E/Trickle Park Res	GU	DEL	34.0	9960	D 3	CUT
CHAIR MOUNTAIN LAKE	N FK GUNNISON R, NE/Paonia Res	GU	GUN	5.7	11505	D 4	CUT
CHALK CREEK LAKE	SW of Mt Princeton Hot Spgs	AR	CHA	3.0	8800	E 6	RBT
CHAPMAN DAM	FRYINGPAN R, SE of Norrie CG	CR	PIT	25.0	8560	D 5	RBT, BRK
CHAPMAN LAKE	FRYINGPAN R, W of Nest	CR	PIT	23.0	9850	D 5	CUT, BRK
CHAPMAN POND	E FK OAK CR, S of Chapman Res	YP	RBL	5.0	9330	B 4	RBT
CHAPMAN RESERVOIR	EAST FORK, OAK CR. NW/Yampa	YP	RBL	25.0	9280	B 4	RBT
CHATFIELD RESERVOIR	S HUNT CR, SW of Yampa	YP	RBL	26.4	10480	B 4	CUT
CHERRY LAKE	GYPSUM CR, EAGLE R, S/Gypsum	CR	EAG	4.2	10660	C 5	CUT
CHIPETA LAKE	UNCOMPAHGRE R, S of Montrose	GU	MON	8.5	5900	F 3	RBT
CHOPPER LAKE	NO NAME CR, ANIMAS R #3	SJ	SNJ	4.0	12620	G 3	CUT
CHRISTINE LAKE	TONER CR, Basalt SWA	CR	EAG	3.0	6590	D 5	RBT, LOC
CIMARRON MGT AREA PONDS	CIMARRON R, S of Cimarron	GU	MON	6.0	8400	F 3	BRK

Name	Location	Drainage	County	Size	Elev.	Map Coor.	Fish Likely to be Present
CLAIRE LAKE	HDWTRS TEXAS CR, E/Taylor Res	GU	GUN	3.6	12360	E 6	CUT
CLEAR CREEK RES	S of Granite	AR	CHA	407.0	8875	D 6	KOK, RBT, LOC, CUT
CLEAR LAKE #2 BEEFPASTUR	CIMARRON R, NE/Silverjack Res	GU	GUN	5.0	9100	F 3	RBT
CLEAR LAKE	DIRTY GEORGE CR, SW Island Lk	GU	DEL	5.4	10180	D 2	CUT
CLEAR LAKE	S FK MINERAL CR, W/Silverton	SJ	SNJ	33.0	12000	G 3	RBT, CUT, BRK
CLEVELAND LAKE	FRENCH CR, NW of Gold Park	CR	EAG	12.9	11820	D 6	CUT, BRK, RBT
CLIFF LAKE	BLACK CR, SW/Green Mtn Res	CR	SUM	8.6	12146	C 6	CUT
CLIFF LAKE	NINEGAR CR, SW of Two Ledge Res	NP	JAC	4.5	9790	B 6	RBT
CLIFF LAKE	S FK WHITE R, E of Hiner Spgs	WR	GAR	3.0	9650	C 4	RBT, BRK
COLBY HORSE PARK RES	LEON CR, PLATEAU CR, S Vega Res	CR	MES	56.2	10120	D 3	CUT, BRK
COLD SPRINGS POND	BEAR R, NE of Stillwater Res	YP	GAR	1.0	10400	B 4	RBT
COLE RESERVOIR	Grand Mesa	CR	MES	8.0	10420	D 3	RBT
COLUMBINE LAKE	JOHNSON CR, SW of Hazel Lk	SJ	LAP	5.2	12370	G 3	CUT
COLUMBINE LAKE	MILL CR, MINERAL CR, NW/Silverton	SJ	SNJ	21.7	12685	G 3	CUT, BRK
COLUMBINE PONDS	Unc Plateau, Columbine Pass	GU	MON	9.0	9000	F 2	RBT
CONFLUENCE LAKE	UNCOMPAHGRE R, NW side of Delta	GU	DEL	18.0	4920	E 2	RBT
CONNECTED LAKES (3)	COLO RIVER & Dike Rd, State Park	CR	MES	———	4620	D 2	LMB, BGL, CRA, CCF
CONSTANTINE LAKE	FALL CR, EAGLE R, SW/Red Cliff	CR	EAG	13.0	11371	D 6	CUT, RBT, BRK
COONEY LAKE	MOSQUITO CR, W of Alma	SP	PAR	8.0	12600	D 6	CUT
COOPER LAKE	COOPER CR, SW of Lake City	GU	HIN	9.6	12750	G 4	CUT
COPPER LAKE, LOWER	COPPER CR, EAST R, NE of Gothic	GU	GUN	7.8	11321	D 5	CUT
COPPER LAKE, UPPER	COPPER CR, EAST R, NE of Gothic	GU	GUN	3.3	12116	D 5	CUT
COPPER MOUNTAIN POND	TEN MILE CR, BLUE RIVER	CR	SUM	6.0	9670	C 6	RBT
CORN LAKE	COLO RIVER & 32 Rd in Clifton, State P	CR	MES	10.0	4620	D 2	RBT
COSTO LAKE	CASTLE CR, OHIO CR, W/Baldwin	GU	GUN	5.7	10980	E 4	BRK
COSTO LAKE, LITTLE	N CASTLE CR, OHIO CR, W/Baldwin	GU	GUN	1.0	10980	E 4	BRK
COTTONWOOD LAKE #1	COTTONWOOD CR, S of Collbran	CR	MES	77.4	10065	D 3	SPL, RBT, BRK, CUT
COTTONWOOD LAKE #4	COTTONWOOD CR, S of Collbran	CR	MES	32.1	10205	D 3	RBT
COTTONWOOD LAKE #5	COTTONWOOD CR, S of Collbran	CR	MES	20.0	9966	D 3	CUT, BRK
COTTONWOOD LAKE	ANIMAS R, E of Hermosa	SJ	LAP	1.7	6640	H 3	CUT
COTTONWOOD LAKE	S COTTWD CR, SW of Buena Vista	AR	CHA	43.0	9552	E 6	RBT, CUT
COVE LAKE RESERVOIR, LOWER	unnamed, MORAPOS CR, WMS FK Y	YP	RBL	4.0	8190	B 3	RBT
COVE RESERVOIR, UPPER	unnamed, MORAPOS CR, WMS FK Y	YP	RBL	5.5	8300	B 3	RBT
COW LAKE	COW CR, WILLOW CR, E/Taylor Res	GU	GUN	5.1	11400	E 6	CUT, BRK
COWDREY LAKE	MICHIGAN R, NW of Walden	NP	JAC	80.0	7940	A 6	RBT, LOC
CRAIG CITY PONDS	S of Craig on Hwy 394	YP	MOF	10.0	6175	A 3	RBT
CRAIG WASTE WATER POND	Waste water Plant SW of Craig	YP	MOF	2.0	6180	A 3	RBT
CRATER LAKE	COW CR TRIB, SW of Overland Res	GU	DEL	4.0	10115	D 3	CUT
CRATER LAKE	CRATER CR, E FK SAN JUAN R	SJ	ARC	20.2	10901	H 5	CUT
CRATER LAKE	CRATER CR, LIME CR, NE/Electra	SJ	SNJ	6.7	11620	G 3	RBT
CRATER LAKE	PINEY R	CR	EAG	8.8	11300	C 6	BRK, RBT
CRATER LAKE	SAND CR, TROUT CR, S of Sand L	YP	RBL	1.0	10150	B 4	BRK
CRESCENT LAKE	S FK DERBY CR. W of Burns	CR	GAR	38.0	10758	C 4	CUT, MAC
CROSHO LAKE	N FK MID HUNT CR, W of Yampa	YP	RBL	56.0	8900	B 4	CUT, GRA, RBT
CROSS LAKE, WEST #2	W CROSS CR, EAGLE R	CR	EAG	4.8	11790	D 5	CUT
CROSS LAKE, WEST #3	W CROSS CR, EAGLE R	CR	EAG	4.8	11835	D 5	CUT
CROSS LAKE, WEST #4	W CROSS CR, EAGLE R	CR	EAG	3.6	12210	D 5	CUT
CRYSTAL LAKE	CRYSTAL CR, HENSON, W/Lake City	GU	HIN	18.0	11760	F 4	BRK, RBT
CRYSTAL LAKE	CRYSTAL CR, TAYLOR R #2	GU	GUN	3.6	11274	E 5	CUT
CRYSTAL LAKE	MID FK MINERAL CR, ANIMAS R #4	SJ	SNJ	4.3	12055	G 3	CUT
CRYSTAL LAKE	SHALLOW CR, W of Creede	RG	MIN	3.0	11560	G 4	CUT
CRYSTAL LAKE	SW of Leadville	AR	LAK	6.7	9400	D 6	RBT
CRYSTAL LAKE, LOWER #1	CRYSTAL CR, SW/Breckenridge	CR	SUM	5.0	11991	D 6	CUT, RBT
CRYSTAL LAKE, LOWER	RACE CR, BEAVER CR, S F RG	RG	RGR	11.4	11220	H 5	CUT, BRK
CRYSTAL LAKE, UPPER	CRYSTAL LK CR, SW/Twin Lks	AR	CHA	8.3	12475	D 6	CUT, RBT
CRYSTAL LAKE, UPPER	RACE CR, BEAVER CR, S F RG	RG	RGR	20.0	11540	H 5	CUT, BRK
D							
DAVIES RESERVOIR, BIG	OAK CR, W of Cedaredge	GU	DEL	10.0	9100	E 2	BRK
DAVIES RESERVOIR, LITTLE	OAK CR, W of Cedaredge	GU	DEL	13.6	9380	E 2	BRK
DEADMAN LAKE	S FK FRYINGPAN R, S of Nast	CR	PIT	2.2	10920	D 5	CUT
DECAMP RESERVOIR	Grand Mesa	CR	MES	9.0	9760	D 3	BRK
DECKERS LAKE	LONGS GULCH, W TENNESSEE CR	AR	LAK	14.0	11350	D 6	CUT
DEEP CREEK LAKE #2, NO.	DEEP CR, MUDDY CR, NE/Paonia R	GU	GUN	11.4	11060	D 4	CUT
DEEP LAKE	DEEP CR, Flattops, E/Heart Lk	CR	GAR	37.0	10470	C 4	BRK, RBT, MAC
DEEP LAKE	NINEGAR CR, Arapaho Lks Group	NP	JAC	1.0	9540	B 6	RBT
DEEP LAKE	W FK E FK WILLIAMS FK YAMPA R	YP	GAR	23.0	10270	B 4	RBT, BRK
DEEP SLOUGH RESERVOIR	Grand Mesa	CR	MES	48.0	10018	D 3	RBT
DEER CREEK LAKES (4)	SLUMGULLIAN PASS, E/Lake City	GU	HIN	8.0	10520	F 4	RBT, CUT
DEER LAKE	MID FK DERBY CR, N of Island L	CR	GAR	5.0	11130	C 4	CUT
DEER LAKE	N of Rabbit Ears Pass	NP	JAC	3.0	10040	B 5	RBT
DELANEY BUTTE, EAST	W of Walden	NP	JAC	65.2	8113	A 5	RBT, CUT, BRK, LOC
DELANEY BUTTE, NORTH	W of Walden	NP	JAC	163.5	8145	A 5	LOC, RBT
DELANEY BUTTE, SOUTH	W of Walden	NP	JAC	150.0	8115	A 5	CUT, RBT, LOC, BRK,
DELUGE LAKE	DELUGE CR, GORE CR, N/Vail Pass	CR	EAG	5.0	11740	C 6	CUT
DENVER LAKE	HORSESHOE CR, N FK ANIMAS R	SJ	SNJ	0.2	11960	G 3	CUT

Western Colorado Lakes

Name	Location	Drainage	County	Size	Elev.	Map Coor.	Fish Likely to be Present
DEVILS LAKE	CANYON INF BRUSH, NE/Lake City	GU	HIN	43.0	11980	F 4	BRK, CUT
DIANA LAKE	N FK ELK R, NE of Pearl Lk	YP	ROU	9.2	10268	A 5	CUT
DIEMER LAKE	N FK FRYINGPAN R, E of Biglow	CR	PIT	11.5	9519	D 5	RBT, BRK
DILLON RESERVOIR (OLD)	hill above Dillon Res, SW/dam	CR	SUM	5.0	9180	C 6	RBT
DILLON RESERVOIR	BLUE R, I-70 & Hwy 9	CR	SUM	3000.0	9017	C 6	RBT, LOC, BRK, CUT, KOK, ARC
DINES LAKE	E FK WMS FK Y, N of Trappers L	YP	GAR	3.0	10342	B 4	CUT
DINKLE LAKE	W SOPRIS CR, SW of Basalt	CR	PIT	9.0	8580	D 4	RBT, LOC
DINOSAUR LAKE	N FK FISH CR, NE/Steamboat Spg	YP	ROU	9.0	10182	A 5	CUT, BRK
DISAPPOINTMENT LAKE	NINEGAR CR, Arapaho Lks Group	NP	JAC	1.5	9700	B 6	RBT
DIVIDE LAKE	LOS PINOS R HDWTRS	SJ	HIN	11.8	9950	G 4	RBT, CUT
DIVIDE LAKE	N FK LAKE CR, W of Twin Lks	AR	LAK	3.2	12378	D 5	CUT
DOG FISH RESERVOIR	Grand Mesa	CR	MES	38.0	10440	D 3	RBT
DOLLAR LAKE	LAKE CR, LOS PINOS, W/Emerald	SJ	HIN	3.0	11560	G 4	CUT
DOLLAR LAKE	RUBY ANTHRACITE CR, SE/Paonia R	GU	GUN	5.0	10023	E 4	BRK
DOLORES FISH UNIT PONDS	old CDOW Fish Unit	DO	MTZ	2.0	7060	H 2	RBT
DOME LAKE	S FK ELK R, E of Clark	YP	ROU	14.7	10060	A 5	CUT
DOME LAKE, UPPER	COCHETOPA CR, S of Gunnison	GU	SAG	75.0	9017	F 5	BRK, RBT
DONUT LAKE	VALLECITO CR HDWTS, LOS PINOS	SJ	SNJ	2.6	12320	G 3	CUT
DORA LAKE	OTTER CR, Green Mtn Res	CR	SUM	2.8	12284	C 6	BRK, CUT
DORIS LAKE	BIG FISH CR, N FK WHITE R	WR	GAR	5.5	10010	B 4	RBT
DOUBLE BUBBLE LAKE	BLACK CR, Green Mtn Res	CR	SUM	25.8	11850	C 6	CUT
DOUGHSPOON RESERVOIRS	DOUGHSPOON CR, Grand Mesa	GU	DEL	4.2	9380	E 2	RBT, BRK
DOUGHTY RESERVOIR	Grand Mesa	CR	MES	24.0	9750	D 3	RBT, CUT
DUGGER RESERVOIR	OAK CR, TONGUE CR, Grand Mesa	GU	DEL	17.0	9180	E 2	BRK, RBT
DUMONT LAKE	NW of Rabbit Earn Pass	CR	GRA	35.0	9508	B 5	BRK
DUNKLEY (DUBEAU) RES	WILLOW CR, E FK WMS FK YAMPA R	YP	ROU	3.3	8950	B 4	RBT
E							
EAGLESMERE LK #1, LOWER	ELLIOTT CR, SW/Green Mtn Res	CR	SUM	4.8	10390	C 6	RBI, CUT, BRK
EAGLESMERE LK #2, UPPER	ELLIOTT CR, SW/Green Mtn Res	CR	SUM	14.4	10397	C 6	RBT, CUT, BRK
EAGLESMERE LK #4	ELLIOTT CR, SW/Green Mtn Res	CR	SUM	14.4	10830	C 6	RBT, CUT, BRK
ECHO CANYON RESERVOIR	ECHO CANYON, S of Pagosa	SJ	ARC	118.0	7100	H 5	RBT, CCF
EDGE LAKE	N FK DERBY CR, NW of Burns	CR	GAR	8.5	10910	C 4	CUT, RBT
EDWARD LAKE	MID FK MAD CR, NE of Mad Creek	YP	ROU	14.7	9867	A 5	RBT
EGGLESTON LAKE, BIG	Grand Mesa, E of Island Lk	GU	DEL	163.8	10140	D 3	RBT
EGGLESTON LAKE, LITTLE	KISER CR, NE of Island Lk	GU	DEL	41.7	10360	D 3	RBT, BRK
EILEEN LAKE	S FK BIG CR, SW of Big Cr Lks	NP	JAC	3.9	10207	A 5	CUT
ELBERT LAKE	S FK MAD CR, NE of Mad Creek	YP	ROU	11.0	10800	A 5	CUT
ELDORADO LAKE, LITTLE	ELK CR, ANIMAS R #3	SJ	SNJ	3.0	12506	G 3	RBT
ELDORADO LAKE, BIG	ELK CR, ANIMAS R #3	SJ	SNJ	14.2	12503	G 3	RBT
ELK LAKE	CANON PASO, W of Granite Lk	SJ	HIN	4.9	11470	G 4	CUT, RBT
ELK PARK LAKE (RES)	W SURFACE CR, NW of Knox Res	GU	DEL	21.2	10065	D 3	RBT, BRK
ELMO LAKE	FISHHOOK CR, SE/Steamboat Spgs	YP	ROU	10.0	10038	B 5	BRK
EMERALD LAKE	EAST R HDWTRS, NW of Gothic	GU	GUN	12.1	10455	D 4	RBT, CUT
EMERALD LAKE	HALFMOON CR, SW of Leadville	AR	LAK	6.0	10000	D 6	RBT
EMERALD LAKE	NEEDLE CR, ANIMAS R, E/Electra	SJ	LAP	8.7	11276	G 3	CUT, RBT
EMMA LAKE	BUCKSKIN CR, NW of Alma	SP	PAR	9.0	12624	D 6	CUT, RBT
EMPEDRADO LAKE	JAROSA CR, W of LaJara Res	RG	CON	36.8	10940	H 6	CUT
ENOCHS LAKE	Pinlon Mesa, SW of Grand Jct	CR	MES	19.0	8910	E 1	RBT
F							
FAIRVIEW LAKE	LOTTIS CR, SE of Taylor Res	GU	GUN	8.7	11296	E 5	BRK
FARWELL LAKE	HINMAN CR, NE of Pear Lk	YP	ROU	4.6	10175	A 5	CUT
FINGER ROCK SETTLING POND	BRINKER CR, CHIMNEY CR	YP	ROU	0.6	8040	B 5	RBT
FINNEY CUT LAKE #1	LEON CR, PLATEAU CR, S Vega Re	CR	MES	3.0	10420	D 3	CUT
FINNEY CUT LAKE #2	LEON CR, PLATEAU CR, S Vega Re	CR	MES	8.5	10490	D 3	CUT
FISH CREEK RESERVOIR	MID FK FISH CR, E/Steamboat Sp	YP	ROU	78.6	9858	B 5	BRK
FISH LAKE	FISH CR, RIO BLANCO, E/Pagosa	SJ	ARC	11.6	11860	H 5	CUT
FLAPJACK LAKE #1	CATARACT CR, SW/Green Mtn Res	CR	SUM	3.0	10710	C 6	CUT
FLAPJACK LAKE #2	CATARACT CR, SW/Green Mtn Res	CR	SUM	2.0	10750	C 6	CUT
FLAPJACK LAKE #3	CATARACT CR, SW/Green Mtn Res	CR	SUM	10.0	10790	C 6	CUT
FLAT LAKE	NINEGAR CR, SE/Seymour Lk SWA	NP	JAC	5.5	9286	B 6	RBT
FLINT LAKE, BIG	FLINT CR, LOS PINOS, N/Emerald	SJ	HIN	30.6	11620	G 4	CUT
FOOSES LAKE, LOWER	W of Poncha Springs	AR	CHA	4.5	8920	E 6	RBT, CUT
FOOSES LAKE, SOUTH	S FOOSES CR HDWTRS, SE/Garfield	AR	CHA	2.7	11400	E 6	CUT
FORBAY RESERVOIR (LAKE)	ELBERT CR, SE of Haviland Res	SJ	LAP	2.5	8300	G 3	BRK
FOREST LAKE	WARD CR, NE of Island Lk	GU	DEL	32.1	10360	D 3	CUT, RBT, BRK
FORTY ACRE LAKE	BIG CR, PLATEAU CR	CR	MES	10.7	10160	D 3	BRK
FRANTZ LAKE	Mt Shavano SFH, NW of Salida	AR	CHA	10.0	7130	E 6	RBT, CUT
FREEMAN RESERVOIR	LIT CTTNWD CR, FORTIFICATION	CR	MOF	16.6	8740	A 4	CUT
FROSTY LAKE	E MARVINE CR, N FK WHITE R	WR	RBL	5.3	9850	B 3	BRK
FRUITA RESERVOIR #1	Pinion Mesa, SW of Grand Jct	CR	MES	11.0	9160	E 1	RBT
FRUITA RESERVOIR #2	Pinion Mesa, SW of Grand Jct	CR	MES	16.0	8820	E 1	RBT
FRUITA RESERVOIR #3	Pinion Mesa, SW of Grand Jct	CR	MES	5.5	8980	E 1	RBT
FULFORD CAVE LAKE	E BRUSH CR, EAGLE R, S/Fulford	CR	EAG	3.0	9200	D 5	RBT
FULLERS LAKE	S MINERAL CR, W of Silverton	SJ	SNJ	19.8	12588	G 3	CUT

Name	Location	Drainage	County	Size	Elev.	Map Coor.	Fish Likely to be Present
G							
GALENA LAKE	BEAR CR, N of Turquoise Lk	AR	LAK	1.8	11050	D 6	CUT
GARDNER PARK RES	BEAR R, SW of Yampa	YP	ROU	47.0	9630	B 4	RBT, LOC
GARFIELD LAKE, LOWER	ELK CR, ANIMAS R #3	SJ	SNJ	5.8	11510	G 3	RBT
GARFIELD LAKE, UPPER	ELK CR, ANIMAS R #3	SJ	SNJ	10.8	12300	G 3	RBT
GILL RESERVOIR	E FK WMS FK YAMPA, SW/Dunkley	YP	ROU	8.0	9020	B 4	BRK
GILLEY LAKE	N ELK CR, S of Buford	WR	RBL	4.7	8305	C 3	BRK
GLACIER LAKE	S FK CONEJOS R, SW of Platoro	RG	CON	21.2	11960	H 5	RBT
GLACIER SPRINGS RET POND	Mesa Lakes Resort, Grand Mesa	CR	MES	3.0	9720	D 2	RBT, BRK
GODDARD LAKE	GODDARD CR, ROUBIDEAU CR	GU	MON	15.0	9240	F 2	RBT
GOLDDUST LAKE #2	E LAKE CR, EAGLE R, S of Avon	CR	EAG	10.0	11558	D 5	CUT
GOLDDUST LAKE #3	E LAKE CR, EAGLE R, S of Avon	CR	EAG	25.0	11916	D 5	CUT
GOLDDUST LAKE #4	E LAKE CR, EAGLE R, S of Avon	CR	EAG	4.8	11740	D 5	CUT
GOLDDUST LAKE #5	E LAKE CR, EAGLE R, S of Avon	CR	EAG	2.0	12070	D 5	CUT
GOODENOUGH RESERVOIR	LEROUX CR, NW of Paonia	GU	DEL	53.4	10500	D 3	CUT, RBT
GOOSE LAKE	FISHER CR, GOOSE CR	RG	MIN	26.7	12000	G 5	CUT
GORE LAKE #1	GORE CR, EAGLE R, N/Vail Pass	CR	EAG	5.0	11400	C 6	CUT
GORE LAKE #2	GORE CR, EAGLE R, N/Vail Pass	CR	EAG	3.0	11975	C 6	CUT
GRANBY RES #1	DIRTY GEORGE CR, NW Cedaredge	GU	DEL	25.4	10080	E 2	RBT
GRANBY RES #12	DIRTY GEORGE CR, NW/Cedaredge	GU	DEL	48.4	10000	E 2	RBT
GRANBY RES #2	DIRTY GEORGE CR, NW/Cedaredge	GU	DEL	18.3	10060	E 2	RBT
GRANBY RES #4,5,10,11	DIRTY GEORGE CR, NW/Cedaredge	GU	DEL	47.0	10000	E 2	RBT
GRANBY RES #7	DIRTY GEORGE CR, NW/Cedaredge	GU	DEL	14.8	9920	E 2	RBT
GRANITE LAKE, LOWER	LOS PINOS R HDWTRS	SJ	HIN	29.5	10780	G 2	RBT
GRASS LAKE	S ARKANSAS R, W of Monarch	AR	CHA	6.1	11500	E 6	CUT, BRK
GREEN LAKE	ANTHRACITE CR, NW of Lk Irwin	GU	GUN	8.0	11635	E 4	CUT
GREEN LAKE	Canon Verde Hdwtrs, S FK CONEJOS	RG	CON	22.7	11600	H 5	CUT, RBT
GREEN LAKE	WILDCAT CR, SW of Crested Butte	GU	GUN	2.7	10613	E 4	CUT
GREEN MOUNTAIN RES.	BLUE R, NW of Dillon	CR	GRA	2125.0	7947	C 6	RBT, LOC, BRK, KOK, MAC
GRIFFITH LAKE	COON CR, PLATEAU CR, Grand Mesa	CR	MES	45.0	10050	D 2	RBT
GRIFFITH LAKE. MIDDLE	COON CR, PLATEAU CR, Grand Mesa	CR	MES	57.0	10025	D 2	BRK, CUT
GRIZZLY GULCH LAKE, LOW	JOHNSON CR, NE of Hazel Lk	SJ	LAP	5.6	12235	G 3	CUT
GRIZZLY LAKE	GRIZZLY CR, N of Duck Lk	CR	GAR	15.0	10510	C 4	BRK
GRIZZLY LAKE	GRIZZLY CR, SW/Independence Pass	CR	PIT	8.0	12510	D 5	RBT
GRIZZLY LAKE	GRIZZLY GUL, CHALK CR, SE/St Elm	AR	CHA	36.3	11202	E 6	SPL, BRK
GRIZZLY RESERVOIR	LINCOLN CR, SE of Aspen	CR	PIT	27.5	11537	D 5	RBT
GROUNDHOG RESERVOIR	GROUNDHOG CR, NE of Dolores	DO	DOL	667.6	8720	G 2	RBT, CUT, LOC, BRK
GYPSUM PONDS	I-70, 1 mile E of Gypsum	CR	EAG	10.0	6410	C 5	RBT, CUT
H							
HACK LAKE	HACK CR, SWEETWATER CR	CR	GAR	2.0	9900	C 4	CUT
HAGERMAN LAKE	BUSK CR, SW of Turquoise Lk	AR	LAK	4.2	11400	D 6	CUT
HAHNS PEAK LAKE	WILSON CR, N of Steamboat Lk	YP	ROU	40.0	8387	A 5	RBT, CUT
HALEY RESERVOIR	BUNKER CR, E Fk Wms F YAMPA R	YP	RBL	12.8	8788	B 4	BRK
HALFMOON LAKE	HALFMOON CR, SW of Leadville	AR	LAK	6.0	11900	D 6	CUT, BRK
HALLENBECK RESERVOIR	N FK KANNAH CR, SE/Whitewater	GU	MES	60.0	5634	E 2	RBT
HANCOCK LAKE, LOWER	CHALK CR HDWTRS, NW/Garfield	AR	CHA	23.0	11615	E 6	CUT
HANCOCK LAKE, UPPER	CHALK CR HDWTRS, NW/Garfield	AR	CHA	7.0	11675	E 6	CUT
HANSON RESERVOIR	Grand Mesa	CR	MES	24.0	9750	D 3	RBT, CUT
HARDSCRABBLE LAKE	E SOPRIS CR, W/Snowmass Village	CR	PIT	2.4	10130	D 4	CUT
HARPER RESERVOIR	EGERIA CR HDWTRS, SW of Topona	YP	ROU	10.0	9820	B 5	BRK
HARRIS LAKE	BIG BRANCH CR, NE of Chromo	SJ	ARC	49.4	9460	H 5	BRK, RBT, CUT
HARRISON FLATS L, LOWER	S FK CLEAR CR, SW of Winfield	AR	CHA	5.0	11909	E 6	CUT
HARTENSTEIN LAKE	DENNY CR, COTTONWOOD CR	AR	CHA	15.0	11432	B 6	CUT
HARVARD LAKE, LOWER	THREE ELK CR, NW of Buena Vista	AR	CHA	2.1	10175	E 6	CUT
HARVEY GAP RES	N of Silt	CR	GAR	196.0	6402	C 3	RBT, SPL, BRK, LOC, LMB,CCF, CRA
HARVEY LAKE	CROSS CR, EAGLE R, SW/Minturn	CR	EAG	20.0	11025	D 5	BRK, CUT
HAVILAND LAKE	ELBERT CR, N of Derange	SJ	LAP	22.0	8106	G 3	RBT, CUT
HAY LAKE	NW of Lake City on FS Trail #235	GU	HIN	2.0	11040	F 4	BRK
HAZEL LAKE	JOHNSON CR, VALLECITO CR #2	SJ	LAP	23.6	12435	G 3	CUT
HEART LAKE	AVALANCHE CR, CRYSTAL R	CR	PIT	2.0	12020	D 4	BRK, RBT
HEART LAKE	DEEP CR, N of Glenwood Spgs	CR	GAR	480.0	10706	C 4	BRK, RBT
HEART LAKE	WATSON CR, SW of Yampa	YP	RBL	48.0	9947	B 4	RBT
HENDERSON LAKE	CANYON CR, NE or Durango	SJ	LAP	14.8	9870	G 3	RBI, BRK
HENRY LAKE	S LOTTIS CR, S of Taylor Res	GU	GUN	12.7	11720	E 5	CUT, RBT
HIDDEN LAKE	CROSBY CR, SW of Coalmont	NP	JAC	10.0	8855	A 5	RBT, CUT
HIDDEN LAKE	IVANHOE CR, FRYING PAN R	CR	PIT	5.O	11550	D 5	CUT
HIDDEN LAKE	MID POWDERHORN CR, Cebola Cr	GU	HIN	1.7	11440	F 4	CUT
HIDDEN LAKE	ROELL CR, VALLECITO CR #2	SJ	LAP	15.9	11940	G 3	CUT, RBT
HIGHLAND MARY LAKE, BIG	CUNNINGHAM CR, ANIMAS R#4	SJ	SNU	42.1	12089	G 3	RBT, CUT, BRK
HIGHLAND MARY LAKE, LIT	CUNNINGHAM CR, ANIMAS R#4	SJ	SNU	11.0	12099	G 3	RBT, CUT, BRK
HIGHLINE LAKE	MACK WASH, NW of Fruita	CR	MES	135.0	4897	D 1	RBT, LMB, WAL, CCF, CRA
HOHNHOLZ LK #1, EAST	LARAMIE R, NW of Glendevey	NP	LAR	40.0	7904	A 6	RBT
HOHNHOLZ LK #2, LITTLE	LARAMIE R, NW of Glendevey	NP	LAR	40.0	7904	A 6	RBT, CUT
HOHNHOLZ LK #3, BIG	LARAMIE R, NW of Glendevey	NP	LAR	40.0	7904	A 6	LOC, CUT
HOLLENBECK PONDS	CORSKE CR, N/Mt Elbert Forebay	AR	LAK	5.0	8450	D 6	CUT, BGL, LMB

Western Colorado Lakes

Name	Location	Drainage	County	Size	Elev.	Map Coor.	Fish Likely to be Present
HOME LAKE	E of Monte Vista	RO	RGR	69.9	7624	G 6	RBT, CCF, NPK
HOME STAKE LAKE, UPPER	HOMESTAKE CR, EAGLE R	CR	PIT	23.3	1092S	D 6	CUT
HOMESTAKE RESERVOIR	HOMESTAKE CR, SW of Red Cliff	CR	EAG	210.0	10260	D 6	CUT, RBT, BRK
HOOPER LAKE #1	N FK DERBY CR, NW of Burns	CR	GAR	22.0	10864	C 4	CUT
HOPE LAKE	LK FK SAN MIGUEL R HDWTRS	DO	SNM	36.6	11900	G 3	CUT
HORSESHOE LAKE	HORSESHOE CR, N FK ANIMAS R	SJ	SNJ	5.9	12536	G 3	CUT
HORSETHIEF LAKE	TRAIL CR, TAYLOR R #2	GU	GUN	3.8	11250	E 5	BRK
HOSSICK LAKE	HOSSICK CR, WEMINUCHE CR	SJ	HIN	9.5	11886	G 4	CUT
HOT SPRINGS RESERVOIR	NE of Doyleville	GU	GUN	43.2	8761	E 5	RBT
HOTEL TWIN LAKE	WARD CR, E of Island Lk	GU	DEL	40.3	10208	D 3	RBT, BRK
HOURGLASS LAKE	RED CANYON, W/Delaney Butte Lk	NP	JAC	7.2	10440	A 5	BRK
HUMPREYS LAKE	ROARING FORK CR, SE of Creede	RG	MIN	45.0	9300	F 5	RBT
HUNKY DORY LAKE, UPPER	N FK SO ARKANSAS R, N/Garfield	AR	CHA	5.0	12040	E 6	CUT
HUNKYDORY LAKE	N FK S ARK R, N of Garfield	AR	CHA	5.0	11860	E 6	CUT, BRK
HUNT LAKE	N FK SO ARKANSAS R, W/Garfield	AR	CHA	6.0	11500	E 6	CUT
HUNTER RESERVOIR	E LEON CR, PLATEAU CR	CR	MES	21.1	10350	D 3	CUT
HUNTERS LAKE	LAKE CR, S FK RIO GRANDE R	RG	MIN	10.0	11383	G 5	BRK, CUT
I							
ILLINOIS LAKE	ILLINOIS CR, TAYLOR R #2	GU	GUN	2.0	10750	E 6	CUT
INDEPENDENCE LAKE	INDEPENDENCE GUL, NW/Lake City	GU	HIN	5.0	11900	F 4	BRK
IRVING LAKE	IRVING CR, VALLECITO CR #2	SJ	LAP	6.3	11662	G 3	CUT
IRWIN LAKE	ANTARACITE CR, W/Crested Butte	GU	GUN	103.5	10323	E 4	RBT, CUT
ISABELLE LAKE	LIT EVANS GUL, NE/Leadville	AR	LAK	2.0	11720	D 6	CUT
ISLAND ACRES LAKE	I-70 NE of Palisade , Colo. R State P	CR	MES	6.0	4758	D 2	RBT
ISLAND LAKE	S MINERAL CR, W of Silverton	SJ	SNJ	8.0	12398	G 3	RBT, BRK, CUT
ISLAND LAKE	WARD CR, Grand Mesa, N/Cedaredge	GU	DEL	177.1	10290	D 2	RBT, BRK, SPL
ISLAND LAKE, LOWER	MID FK DERBY CR, W of Burns	CR	GAR	28.0	10866	C 4	CUT, RBT
ISLAND LAKE, MIDDLE	MID FK DERBY CR, W of Burns	CR	GAR	15.0	11180	C 4	CUT
ISLAND LAKE, UPPER	MID FK DERBY CR, W of Bums	CR	GAR	27.0	11202	C 4	CUT
ISLANDLAKE	N FK S ARKANSAS R, NW of Garfield	AR	CHA	8.7	11800	E 6	CUT
ISLANDLAKE	NearTown of Marble	CR	GUN	2.0	7770	D 4	RBT
ISOLATION LAKE #1	E FK HOMESTAKE CR, EAGLE R	CR	PIT	5.0	11580	D 6	CUT
ISOLATION LAKE #2	E FK HOMESTAKE CR, EAGLE R	CR	PIT	5.0	11620	D 6	CUT
ITALIAN LAKE	ITALIAN CR, TAYLOR R #2	GU	GUN	5.7	11675	E 5	CUT
IVANHOE LAKE	IVANHOE CR, FRYINGPAN, SE/Nast	CR	PIT	72.6	10929	D 5	RBT
J							
J O K RES #1	W FK FISH CR, S of Dunckley	YP	ROU	2.9	9350	B 4	RBT
JACK LAKE	LINCOLN CR, E of Truro Lk	CR	PIT	2.4	12240	D 5	CUT
JACKSON GULCH RES.	N of Mancos	SJ	MTZ	217.0	7825	H 2	RBT, CCF
JACOBS LADDER LAKE	FIRST FK PIEDRA R, SE/Red Cr	SJ	ARC	12.0	8640	H 4	BRK
JET LAKE	PATTERSON CR, S FK WHITE R	WR	GAR	15.0	10335	C 4	RBT, BRK
JEWEL LAKE	NEEDLE CR, ANIMAS R #3	SJ	LAP	8.7	11910	G 2	CUT
JOE MOORE RESERVOIR	NW of Mancos	SJ	MTZ	33.0	7600	H 2	RBT, LOC, LMB
JOHNSON LAKE	E MARVINE CR, N FK WHITE R	WR	RBL	2.5	9020	C 4	RBT
JOHNSON LAKE	JOHNSON CR, VALLECITO CR #2	SJ	LAP	4.0	12100	G 3	CUT
JONAH LAKE	CHEDSEY CR, NE of Buffalo Pass	NP	JAC	8.6	10164	A 5	CUT, BRK, RBT
JUMBO RESERVOIR	Mesa Lakes Resort, Grand Mesa	CR	MES	4.5	9780	D 2	RBT, BRK
JUMPER LAKE	JUMPER CR, TROUT CR, RG R #4	RG	MIN	8.4	11577	G 4	CUT
JUNIATA RESERVOIR	KANNAH CR, SE of Whitewater	GU	MES	160.0	5716	E 2	RBT
K							
KATHLEEN LAKE	COYOTE CR, S of Seymour Res	NP	JAC	2.5	9500	B 6	CUT
KATOS POND	GORE CR, EAGLE R	CR	EAG	1.0	8300	C 6	RBT
KEENER LAKE	N FK DERBY CR, NW of Burns	CR	GAR	24.0	10780	C 4	RBT
KENNEY RESERVOIR	WHITE R #1, E of Rangley	WR	RBL	600.0	5350	B 1	RBT
KENNY CREEK RESERVOIR	LEON CR, PLATEAU CR, S/Vega Res	CR	MES	17.3	9880	D 3	RBT, CUT, BRK
KERR LAKE	ALAMOSA R, NE of Platoro	RG	CON	39.5	11380	H 5	CUT
KIDNEY LAKE	NINEGAR CR, Arapaho Lks Group	NP	JAC	4.3	9018	B 6	RBT
KISER SLOUGH RESERVOIR	WARD CR, S of Reed Res	GU	DEL	34.1	9827	D 3	RBT
KITE LAKE	BUCKSKIN CR, NW of Alma	AR	PAR	7.0	12033	D 6	CUT, RBT
KITSON RESERVOIR	COTTONWOOD CR, PLATEAU CR	CR	MES	13.0	10040	D 3	RBT, BRK, GRA
KLINE'S FOLLY	Flattops near Deep Lk	CR	GAR	1.0	10710	C 4	RBT
KNOX RESERVOIR	Grande Mesa	CR	MES	16.0	10000	C 3	BRK, RBT
KROENKE LAKE	N COTTONWOOD CR, NW/Buena Vista	AR	CHA	24.0	11600	B 6	CUT
L							
L.E.D.E. RESERVOIR	GYPSUM CR, l5 mi SE of Gypsum	CR	EAG	27.3	9530	D 5	RBT
LA JARA RESERVOIR	LA JARA CR, W of LaJara	RG	CON	1375.0	9698	H 6	BRK
LAKE JOHN	NW of Walden	NP	JAC	656.0	8048	A 5	RBT, CUT, LOC, BRK
LAKE OF THE CRAGS	N FK MAD CR, NE of Mad Creek	YP	ROU	4.8	10850	A 5	CUT
LAKE OF THE WOODS	N FK WHITE R, NW Trappers Lk	WR	GAR	8.0	10000	B 4	BRK, CUT
LAME DUCK LAKE	BLACK CR, SW Green Mtn Res	CR	SUM	2.0	11890	C 6	CUT
LAMPHIER LAKE, LOWER	LAMPHIER CR, GOLD CR, NW/Pitkin	GU	GUN	3.5	11270	E 5	CUT
LAMPHIER LAKE, UPPER	LAMPHIER CR, GOLD CR, NW/Pitkin	GU	GUN	5.8	11770	E 5	CUT
LANNING LAKE	Grand Mesa	CR	MES	3.0	10310	D 3	BRK, CUT
LARSON LAKE	LARSON CR, NW of Lake City	GU	HIN	3.4	11160	F 4	RBT, BRK
LAVA LAKE	LAVA CR, PINEY, SE/State Bridge	CR	EAG	2.0	9610	C 5	BRK

Name	Location	Drainage	County	Size	Elev.	Map Coor.	Fish Likely to be Present
LE PLATT LAKE	Bayfield	SJ	LAP	10.0	6950	H 3	RBT
LEEMAN LAKES	W BRUSH CR, EAGLE R, S/Sylvan L	CR	EAG	5.0	10130	D 5	CUT
LEMON RESERVOIR	FLORIDA R, NE of Durango	SJ	LAP	622.0	8148	H 3	RBT, KOK
LEON LAKE	Grand Mesa	CR	MES	94.0	10360	D 3	BRK
LEON PEAK RESERVOIR	Grand Mesa	CR	MES	9,0	10640	D 3	CUT
LEVIATHAN LAKE, NORTH	LEVIATHAN CR, VALLECITO CR #2	SJ	SNJ	6.4	12460	G 3	CUT
LEVIATHAN LAKE, SOUTH	LEVIATHAN CR, VALLECITO CR #2	SJ	SNJ	7.7	11978	G 3	CUT
LIGON RESERVOIR	EGERIA CR, 1.5 mi W of Egena	CR	ROU	5.0	8180	B 5	RBT
LILLIE LAKE	FLORIDA R HDWTRS, NCity Res	SJ	LAP	5.7	12550	G 3	CUT
LILY LAKE #1	ROCK CR, PINEY SE/State Bridge	CR	EAG	1.0	9030	C 5	BRK
LILY LAKE	COTTONWOOD CR, PLATEAU CR	CR	MES	4.2	10220	D 3	RBT, BRK
LILY LAKE	LIT MUDDY CR, S/Rabbft Ears Pass	CR	GRA	10.0	9080	B 5	BRK
LILY PAD LAKE	MEADOW CR, W of Dillon Res	CR	SUM	5.0	9964	C 6	BRK
LILY POND	ALAMOSA R, NW of Platoro	RG	CON	24.0	10940	H 5	CUT
LIMESTONE LAKE	PATTERSON CR, W of Elk Lakes	WR	GAR	5.0	10625	C 4	BRK
LINKINS LAKE	LINKINS CR, ROARING FK R	CR	PIT	22.0	12008	D 5	RBT, BRK
LITTLE GEM LAKE	GENEVA CR, N of Galena Lk	CR	GUN	3.0	11660	D 4	CUT
LITTLE GEM RESERVOIR	WARD CR, SE of Island Lk	GU	DEL	39.5	10070	D 2	RBT, BRK. SPL
LITTLE HILLS POND	PICEANCE CR, CDOW land, SW/Meeker	WR	RBL	5.2	6160	B 2	CUT, RBT, BRK
LIZARD LAKE	LOST TRAIL CR, E of Marble	CR	GUN	5+	8600	D 4	BRK
LONE LICK LAKES	LONE LICK CR, E/State Bridge	CR	EAG	4.0	9780	C 5	CUT, BRK
LONESOME LAKE #1	E FK HOMESTAKE CR, EAGLE R	CR	PIT	6.0	11160	D 6	CUT
LONESOME LAKE #2	E FK HOMESTAKE CR, EAGLE R	CR	PIT	10.0	11350	D 6	CUT
LONG LAKE	E FK WMS FK Y N/Trappers Lk	YP	GAR	21.6	10464	B 4	CUT, BRK
LONG LAKE	NINEGAR CR, Arapaho Lks Group	NP	JAC	3.0	9343	B 6	RBT
LONG LAKE	PASS CR, OHIO CR	GU	GUN	3.2	10080	E 4	CUT
LONG LAKE	PINEY R	CR	EAG	7.5	11100	C 6	CUT
LONGS LAKE, LOWER	WILLOW CR, SW of Rand	NP	JAC	2.0	9860	B 6	CUT
LONGS LAKE, UPPER	WILLOW CR, SW of Rand	NP	JAC	2.8	10000	B 6	CUT
LOST LAKE	ELK CR, ANIMAS R #3	SJ	SNJ	10.2	12182	G 3	RBT, BRK
LOST LAKE	FISHHOOK CR, NE of Fishhook Lk	YP	ROU	15.0	9920	B 5	BRK
LOST LAKE	GYPSUM CR, EAGLE R, SW/Sylvan L	CR	EAG	5.1	10860	D 5	CUT
LOST LAKE	JAROSA CR, W of La Jara Res	RG	CON	28.2	10580	H 6	CUT
LOST LAKE	KENNY CR, above Kenny Cr Res	CR	MES	3.6	10240	D 3	CUT, BRK
LOST LAKE	MIDDLE CR, SE of Paonia Res	GU	GUN	11.9	9870	E 4	BRK
LOST LAKE	OFFICERS GULCH, SW/Dillon Res	CR	SUM	2.0	11590	C 6	CUT
LOST LAKE	RED SANDSTONE CR, N of Vail	CR	EAG	14.0	10158	C 6	MAC, RBT
LOST LAKE	ROELL CR, VALLECITO CR #2	SJ	LAP	17.0	11839	G 3	CUT
LOST LAKE SLOUGH	MIDDLE CR, SE of Paonia Res	GU	GUN	76.3	9823	E 4	RBT, BRK
LOST LAKE	TIE CR, E FK SAN JUAN R	SJ	MIN	7.7	8640	H 5	CUT
LOST LAKE	W DIVIDE CR HDWTRS, Quaker Mesa	CR	MES	1.5	10530	D 4	CUT
LOST LAKE	WILLOW CR, SE of South Fork	RG	RGR	1.0	10120	G 5	CUT
LOST LAKE, EAST	W FK E FK WMS FK Y, N/Trappers	YP	GAR	13.8	10300	B 4	CUT, RBT
LOST MAN RESERVOIR	ROARING FK R, W of Linkins Lk	CR	PIT	7.2	10640	D 5	RBT, BRK
LOST SOLAR LAKE #1	LOST SOLAR CR. S FK WHITE R	WR	GAR	4.0	10630	C 4	CUT
LOSTLAKE	unnmd to N FK MID COTTONWOOD C	AR	CHA	4.0	11880	E 6	CUT
LOTTIS LAKE, UPPER	S LOTTIS CR, TAYLOR CR #1	GU	GUN	6.4	11630	E 5	CUT
LYNX PASS RESERVOIR	LIT ROCK CR. NE of Toponas	YP	ROU	2.5	8937	B 5	RBT
M							
MACHIN LAKE	MID FK SAGUACHE CR HDWTRS	RG	SAG	11.4	12580	G 5	CUT
MACK MESA LAKE	10 mi N of Loma	CR	MES	30.0	4730	D 1	RBT, CCF, LMB, WAL
MACKINAW LAKE	S FK DERBY CR, W of Burns	CR	GAR	30.0	10756	C 4	CUT, MAC
MAHAFFEY LAKE	PATTERSON CR, S FK WHITE R	WR	RBL	20.0	9783	C 4	CUT
MAHAN LAKE #1	N FK ELLIOTT CR, SWGrn Mtn Res	CR	SUM	8.2	10814	C 6	BRK
MANDALL LAKE, BLACK	MANDALL CR, BEAR R, SW/Yampa	YP	GAR	9.0	10820	B 4	RBT
MANDALL LAKE, MUD	MANDALL CR, BEAR R, SW/Yampa	YP	GAR	2.0	10550	B 4	BRK
MANDALL LAKE, SLIDE	MANDALL CR, BEAR R, SW/Yampa	YP	GAR	12.0	10670	B 4	CUT
MANDALL LAKE, TWIN UP	MANDALL CR, BEAR R, SW/Yampa	YP	GAR	7.5	10550	B 4	CUT
MARGARET LAKE	MID FK MAD CR, NE of Mad Creek	YP	ROU	33.1	9987	A 5	CUT
MAROON LAKE	W MAROON CR, SW of Aspen	CR	PIT	25.0	9580	D 5	RBT, BRK
MARTIN LAKE	LA PLATA GULCH, LAKE CR	AR	CHA	1.5	11360	D 6	CUT
MARVINE LAKE LAKES	MARVINE CR, SW of Trappers Lk	WR	RBL	88.0	9317	C 4	RBT, BRK
MC CABE POND	Gunnison, S/River E side MC #1	GU	GUN	2.2	7730	E 5	CUT
MC CONOUGH RESERVOIR	BULL CR, SW of Dome Lakes	GU	SAG	54.3	9420	F 5	RBT
MC CULLOUGH LAKE #2	MC CULLOUGH GUL, SW/Breckenridge	CR	SUM	3.6	12550	D 6	CUT
MC CULLOUGH LAKE #4	MC CULLOUGH GUL, SW/Breckenridge	CR	SUM	5.0	12500	D 6	CUT
MC CULLOUGH LAKE #6	MC CULLOUGH GUL, SW/Breckenridge	CR	SUM	5.0	12700	D 6	CUT
MC CURRY RES	E HAUXHURST CR, PLATEAU CR	CR	MES	22.6	10474	D 3	CUT
MC GINNIS LAKE	SKINNY FISH CR, N FK WHITE R	WR	GAR	22.5	10158	B 4	CUT, RBT
MC KEE LAKE	Near Town of Marble	CR	GUN	2.0	8100	D 4	RBT, BRK
MC MILLAN LAKE	MID FK DERBY CR, NW of Burns	CR	GAR	2.0	9238	C 4	CUT
MC PHEE RESERVOIR	DOLORES R #4, NW of Dolores	DO	MTZ	4470.0	6924	G 1	KOK, RBT, BGL, CCF
MEADOW LAKE	MEADOW CR, 15 mi N/New Castle	CR	GAR	60.0	9625	C 4	BRK, RBT
MEADOWS RES, BIG	HELLS HOLE CR, PLATEAU CR	CR	MES	39.S	9790	D 3	RBT
MERIDIAN LAKE	WASHINGTON GULCH, SLATE R	GU	GUN	38.5	9820	E 4	RBT

Western Colorado Lakes

Name	Location	Drainage	County	Size	Elev.	Map Coor.	Fish Likely to be Present
MESA LAKE	Mesa Lakes Resort, Grand Mesa	CR	MES	26.0	9840	D 2	RBT, BRK
MESA LAKE, SOUTH	MESA CR, PLATEAU CR	CR	MES	10.0	9980	D 2	BRK
MICA LAKE	GILPIN CR, ELK R. E/Pearl Lk	YP	ROU	5.5	10428	A 5	CUT
MILITARY PARK RES	W FK SURFACE CR, NW/Trickle Park	GU	DEL	30.6	10100	D 3	RBT, CUT
MILL CREEK PONDS	upper CEBOLLA CR, SE/Lake City	GU	HIN	1.5	10950	G 4	RBT
MILL LAKE	MILL CR, GOLD CR, NW of Pitkin	GU	GUN	5.9	11674	E 5	CUT
MILLIONS LAKE	MILL CR, SW of South Fork	RG	RGR	3.7	8460	G 5	RBT
MIRAMONTE RESERVOIR	NATURITA CR, S of Norwood	DO	SNM	405.0	7700	G 2	RBT
MIRROR LAKE	E WILLOW CR, E of Tincup	GU	GUN	27.4	10962	E 6	RBT, BRK, LOC
MIRROR LAKE	MID FK MAD CR, NE of Mad Creek	YP	ROU	6.4	10040	A 5	CUT
MIRROR LAKE, LOWER	CATARACT CR, GREEN MTN RES	CR	SUM	8.1	10559	C 6	CUT
MISSOURI LAKE #1	MISSOURI CR, HOMESTAKE CR	CR	EAG	4.0	11380	D 5	CUT, BRK
MISSOURI LAKE #2	MISSOURI CR, HOMESTAKE CR	CR	EAG	4.0	11420	D 5	CUT, BRK
MIX LAKE	CONEJOS R, N of Platoro Res	RG	CON	21.7	10080	H 5	RBT
MOHAWK LAKE #2	SPRUCE CR, SW of Breckenridge	CR	SUM	12.0	12100	D 6	CUT
MOLAS LAKE, BIG	S of Silverton, Molas Pass	SJ	SNJ	20.0	10500	G 3	RBT, BRK
MOLAS LAKE, LITTLE	SW of Silverton, Molas Pass	SJ	SNJ	6.9	10906	G 3	RBT, BRK
MONARCH VALLEY RANCH POND	HICKS CREEK, TOMICHI CREEK	GU	SAG	2.0	8300	F 6	RBT
MONTGOMERY RESERVOIR	MID FK SP R, NW of Fairplay	SP	PAR	97.0	10840	D 6	RBT, CUT
MONUMENT RES #1	MONUMENT CR, LEON CR, Grand Mesa	CR	MES	20.0	10196	D 3	CUT
MOON LAKE	LAKE CR, LOS PINES R	SJ	HIN	14.8	11653	G 4	CUT, RBT
MOON LAKE	W SNOWMASS CR, SW/Snowmass Villa	CR	PIT	9.9	11720	D 4	CUT
MORRIS RESERVOIR	OAK CR, W of Cedaredge	GU	DEL	2.7	9180	E 2	RBT
MORROW POINT RESERVOIR	GUNNISON R, NE of Cimarron	GU	MON	807.0	7160	F 3	CUT, KOK, RBT, LOC
MT ELBERT FORBAY	N of Twin Lake	AR	LAK	200.0	9500	D 6	RBT, CUT
MT GUNNISON LAKE #1, NO.	CASCADE CR, S of Paonia Res	GU	GUN	4.7	11000	E 4	RBT
MT GUNNISON LAKE #2, SO.	CASCADE CR, S of Paonia Res	GU	GUN	11.6	11208	E 4	RBT
MUD LAKE #4	MUSKRAT C, MID DERBY CR	CR	GAR	1.0	10280	C 1	RBT
MUDDY PASS LAKE	GRIZZLY CR, E/Rabbit Ears Pass	NP	JAC	10.5	8780	B 5	RBT
MULHALL LAKE #2	HOMESTAKE CR, NW of Gold Park	CR	EAG	6.3	11980	D 6	RBT, CUT
MURPHY LAKE	MARVINE CR, N FK WHITE R	WR	RBL	4.0	9350	B 4	CUT
MYSTERIOUS LAKE	S ITALIAN CR, N/Spring Cr Res	GU	GUN	6.4	11274	E 5	BRK
MYSTERY LAKE	VALLECITO CR #2, LOS PINOS R	SJ	SNJ	9.3	12420	G 3	CUT
N							
N FK MICHIGAN R LAKE	N FK MICHIGAN R, NE of Gourd	NP	JAC	65.9	9626	A 6	RBT, LOC
N HALFMOON LK, LOWER	N HALFMOON CR, SW/Turquoise Lk	AR	LAK	12.5	12025	D 6	CUT
N HALFMOON LK, MIDDLE	N HALFMOON CR, SW/Turquoise Lk	AR	LAK	11.3	12220	D 6	CUT
NANCY LAKE	CARTER CR, N FK FRYINGPAN R	CR	PIT	7.0	11370	D 5	CUT
NARRAGUINNEP RES	W F DOLORES R, 8 ml E of Dolores	CR	MON	386.0	7050	H 2	RBT, NPK, CRA, CCF, BGL
NAST LAKE	FRYINGPAN R, 5 mi SE of Norrie	CR	PIT	8.0	8710	D 5	RBT
NATIVE LAKE	CHAMA-DIAMONDS RANCH	RG	ARC	5.0	10000	H 5	CUT
NATIVE LAKE	ROCK CREEK, W of Leadville	AR	LAK	5+	11200	D 6	CUT
NAVAJO RESERVOIR	PIEDRA R, E of Ignacio, State Park	CR	ARC	15000	6200	H 4	BGL, CCF, CRA, LMB, KOK, RBT
NED WILSON LAKE	E MARVINE CR, N FK WHITE R	WR	RBL	3.0	11075	C 4	BRK, CUT
NEEDLE CREEK RESERVOIR	NEEDLE CR, TOMICHI CR	GU	SAG	40.0	8830	F 5	BRK, RBT
NEVER SWEAT RESERVOIR	16 mi S of Collbran	CR	MES	19.0	10070	D 3	RBT, BRK
NEW YORK LAKE	NEW YORK CR, LINCOLN CR	CR	PIT	1.9	11950	D 5	CUT
NEW YORK LAKE	W LAKE CR HDWTRS, S/Edwards	CR	EAG	40.0	11274	D 5	CUT
NO NAME LAKE	NO NAME CR, ANIMAS R #3	SJ	SNJ	11.0	12552	G 3	CUT
NO NAME LAKE	S FK CONEJOS R, S of Platoro	RG	CON	39.5	11360	H 5	CUT
NORTH FORK RES	N FK S ARK R, NW of Maysville	AR	CHA	20.0	11420	E 6	CUT, RBT
O							
O'HAVER RESERVOIR	GRAYS CR, W of Poncha Pass	AR	CHA	14.4	9160	F 6	RBT, CUT
OAK LAKE	TROUT CR, S of Sheriff Res	YP	RBL	2.0	10150	B 4	CUT, RBT
OAK RIDGE SWA SEEP	S FK WHITE R, SW of Buford	WR	RBL	1.0	6900	C 3	CUT
OFFICERS GULCH POND	on I-70 2 mi W of Frisco	CR	SUM	3.0	9500	C 6	RBT
OLATHE RESERVOIR #2	E DRY CR, between E & W Dry Crs	GU	MON	10.0	7640	F 2	RBT
OLIVER TWIST LAKE	MOSQUITO CR, W of Alma	SP	PAR	6.0	12150	D 6	CUT
OPAL LAKE	WHITE CR, E of Echo Canyon Res	SJ	ARC	19.8	9220	H 5	RBT
OSH LAKE	Pearl	NP	JAC	3.0	7800	A 5	RBT
OVERLAND RESERVOIR	COW CR, W MUDDY CR, NW/Paonia	GU	DEL	232.0	9897	D 3	RBT, CUT, BRK
P							
PACIFIC LAKE	SPRUCE CR, BLUE R #3	CR	SUM	12.5	13419	D 6	CUT
PAGODA LAKE	SNELL CR, N FK WHITE R	WR	RBL	10.0	10313	B 4	BRK, CUT
PALMER LAKE	E NO NAME CR, NE/Glenwood Sprgs	CR	GAR	10.0	10670	E 4	BRK, CUT
PAONIA	GUNNISON, E Of Paonia, State Park	GU	GUN	334.0	6500	E 4	RBT, NPK
PARADISE LAKE #1	HOMESTAKE CR, EAGLE R	CR	PIT	7.4	11230	D 6	CUT, BRK
PARADISE LAKE #3	HOMESTAKE CR, EAGLE R	CR	PIT	2.4	11560	D 6	CUT, BRK
PARADISE LAKE #4	HOMESTAKE CR, EAGLE R	CR	PIT	4.2	11590	D 6	CUT, BRK
PARK LAKE RES	E DIVIDE CR HDWTRS	CR	MES	5.5	9050	D 4	RBT
PASS CREEK POND	PASS CR, NE of Wolf Cr Pass	RG	MIN	14.8	9240	G 5	RBT
PASTORIUS RESERVOIR	FLORIDA R, SE of Durango	SJ	LAP	53.0	6880	H 3	RBT, NPK, BGL, YPE, CCF, LMB
PATRICIA LAKE	E CROSS CR, EAGLE R, SW/Red Cl	CR	EAG	12.0	11398	D 6	CUT
PATRICIA LAKE	PINEY R, N of Vail	CR	EAG	5.0	10550	C 6	CUT
PEAR LAKE	N TEXAS CR (Mnged by SE Region)	GU	GUN	8.0	11900	E 6	CUT

Name	Location	Drainage	County	Size	Elev.	Map Coor.	Fish Likely to be Present
PEARL LAKE	LESTER CR, E of Steamboat Lk	YP	ROU	167.0	8050	A 5	CUT, RBT, BRK
PEARL LAKE	NEEDLE CR, ANIMAS R #3	SJ	LAP	6.9	11579	G 3	CUT, RBT
PEELER LAKE, LOWER #1	PEELER CR, OH-BE-JOYFUL, SLATE	GU	GUN	5.2	10780	E 4	BRK
PEGGY LAKE	HILL CR, W of Boettcher Lk	NP	JAC	9.8	11165	A 5	CUT
PERICLES POND	below Ridgway dam, South pond	GU	OUR	1.3	6950	F 3	RBT
PETERSON DRAW RESERVOIR	WOLF CR, NE of Massadona	YP	MOF	2.0	5810	B 1	RBT
PETROLEUM LAKE	LINCOLN CR, ROARING FK R	CR	PIT	12.4	12310	D 5	RBT
PIERRE LAKE #1	BEAR CR, SW/Snowmass Village	CR	PIT	4.3	12070	D 4	CUT
PIERRE LAKE #2, LOWER	BEAR CR, SW/Snowmass Village	CR	PIT	12.9	12130	D 4	CUT
PIERRE LAKE #3	BEAR CR, SW/Snowmass Village	CR	PIT	44.3	12190	D 4	CUT
PIERRE LAKE #4, UPPER	BEAR CR, SW/Snowmass Village	CR	PIT	15.7	12340	D 4	CUT
PINE ISLE LAKE	MARVINE CR, N FK WHITE R	WR	RBL	7.0	9230	C 4	RBT, CUT
PINE RESERVOIR	YOUNGS, WARD CR, E/Kiser Res	GU	DEL	5.7	10000	D 3	CUT
PINEY LAKE	PINEY R, N of Vail	CR	EAG	45.0	9342	C 6	RBT, BRK
PITCAIRN RESERVOIR	OAK CREEK, W of Cedaredge	GU	DEL	20+	9400	E 2	BRK
PITKIN LAKE	PITKIN CR, NE of Vail	CR	EAG	8.0	11380	C 6	CUT
PLATORO RESERVOIR	CONEJOS R #4, W of Platoro	RG	CON	700.0	9970	H 5	RBT, KOK, LOC
POAGE LAKE	BEAVER CR, S FK RIO GRANDE R	RG	RGR	28.9	11065	H 5	CUT
POMEROY LAKE, LOWER	POMEROY GUL, CHALK CR, S/St Elm	AR	CHA	35.0	12035	E 6	CUT
POMEROY LAKE, UPPER	POMEROY GUL, CHALK CR, S/St Elm	AR	CHA	37.0	12300	E 6	CUT
PORCUPINE LAKE	S FK MAD CR, NE of Mad Creek	YP	ROU	4.2	9860	A 5	RBT
POT HOLE LAKE #2(UPPER)	Taylor Park, N of Taylor Res	GU	GUN	13.6	9762	E 5	RBT
POTHOLE LAKE #1 (LOWER)	Taylor Park, N of Taylor Res	GU	GUN	4.9	9675	E 5	RBT
POWDERHORN LAKE, LOWER	W FK POWDERHORN CR, Cebolla Cr	GU	HIN	7.4	11650	F 4	BRK
POWDERHORN LAKE, UPPER	W FK POWDERHORN CR, Cebolla Cr	GU	HIN	25.7	11860	F 4	CUT, BRK
PRICE LAKES	NAVAJO R, NE of Chromo	SJ	ARC	4.9	8880	H 5	RBT
PTARMIGAN LAKE	RED MTN CR, S of Camp Bird	GU	OUR	4.9	12939	G 4	CUT
PTARMIGAN LAKE	TELLURIUM CR, HDWTRS Taylor Cr.	GU	GUN	5.0	12300	D 5	RBT, CUT
PTARMIGAN LAKE	WOLVERINE CR, S FK ELK R	YP	ROU	7.3	10699	A 5	CUT
PTARMIGAN LAKE, LOWER	MID COTTONWD CR, W/Buena Vista	AR	CHA	7.0	11760	E 6	CUT
PTARMIGAN LAKES	PTARMIGAN CR, MID CTTNWD CR	AR	CHA	27.9	12147	E 6	CUT
PUETT RESERVOIR	NE of Cortez	DO	MTZ	146.0	7260	H 2	NPK, RBT, WAL
Q							
QUAKER MOUNTAIN LAKE	ELKHEAD CR, SW/Steamboat Lk	YP	ROU	2.0	8160	A 4	RBT
QUARTZ LAKE	QUARTZ CR, E FK SAN JUAN R	SJ	ARC	6.9	11600	H 5	CUT
QUARTZITE LAKE	S FK GRIZZLY CR, N of Glenwood Sp	CR	GAR	5.0	10590	C 4	BRK
R							
RAFT LAKE	Rabbit Ears Pass	NP	JAC	5.7	10120	B 5	CUT
RAINBOW LAKE	CAUSEWAY CR, BEAR R, SW/Yampa	YP	GAR	8.0	10764	B 4	CUT
RAINBOW LAKE	E LAKE CR, S of Edwards	CR	EAG	3.6	11040	C 5	CUT
RAINBOW LAKE	MARVINE CR. N FK WHITE R	WR	RBL	4.0	9520	B 4	RBT, CUT
RAINBOW LAKE	NORRIS CR, SW/Delaney Lks, Rawah	NP	JAC	96.0	9854	A 5	RBT, CUT
RAINBOW LAKE	PINE CR, W of Riverside	AR	CHA	5.0	11650	E 6	CUT
RAINBOW LAKE	TROUT CR, S of Sheriff Res	YP	RBL	1.4	9880	B 4	CUT, RBT
RAINBOW LAKE	WILLOW CR, N of Blue Mesa Res	GU	GUN	11.6	10874	E 4	BRK, RBT
RAINBOW LAKE, BIG	NORRIS CR, SW/Delaney Lks, Rawah	NP	LAR	7.2	10722	A 5	RBT
RAINBOW LAKE, LOW	NORRIS CR, SW/Delaney Lks, Rawah	NP	JAC	9.0	9700	A 5	RBT
RAINBOW LAKE, MID	NORRIS CR, SW/Delaney Lks, Rawah	NP	JAC	9.0	9830	A 5	RBT
RAMS HORN LAKE	LITTLE DOME CR, BEAR R	YP	GAR	8.0	9897	B 4	BRK, RBT
RED MOUNTAIN LAKE	ALKALI CR, NW/Roaring Judy SFU	GU	GUN	1.5	10360	E 4	CUT
RED MTN LAKE, UPPER	TAYLOR R (Managed by SE Region)	GU	GUN	2.0	12040	E 5	CUT
REED RESERVOIR	WARD CR, S of Barron Res	GU	DEL	22.2	9980	D 3	RBT
REGAN LAKE	HOUSE CANYON, CROOKED, SPG CR	RG	HIN	80.0	10041	G 4	BRK, RBT
RIDGWAY RESERVOIR	UNCOMPAGRE R #3B, N/Ridgway	GU	OUR	1000.0	7000	F 3	RBT, CUT
RIFLE GAP RESERVOIR	W RIFLE CR, 10 mi N of Rifle, State Park	CR	GAR	400.0	5950	C 3	CUT, RBT, BRK, WAL, LMB
RIFLE POND, SOUTH	l-70 W of Rifle	CR	GAR	3.0	5315	C 3	RBT
RIM LAKE	RED CR, GYPSUM CR, NW/Ruedi Res	CR	EAG	7.0	10640	D 5	RBT, CUT, BRK
RIM LAKE	SWEETWATER CR, S of Trappers L	CR	GAR	10.0	10804	C 4	CUT, MAC
RIMROCK LAKE	WARD CR, SW of Little Gem Lk	GU	DEL	12.0	10120	D 2	RBT, CUT
RIO BLANCO LAKE	WHITE R #1, NW of Meeker	WR	RBL	116.0	5756	B 2	RBT, LMB, NPK, CCF, CPA
RIO GRANDE RESERVOIR	RIO GRANDE R #5, S/Lake City	RG	HIN	1500.0	9392	G 4	RBT, CUT
RITO HONDO LAKE	RITO HONDO CR, N CLEAR CR	RG	HIN	41.0	10280	G 4	RBT, BRK
RIVERSIDE PONDS	upstrm of Mt Shavano SFU	AR	CHA	3.0	7120	E 6	RBT, CUT
ROAD CANYON RESERVOIR	LONG CANYON, W of Creede	RG	HIN	140.0	9275	G 4	RBT, BRK
ROARING JUDY PONDS	Roaring Judy SFU, N of Almont	GU	GUN	3.0	8040	E 5	RBT
ROCK LAKE #1, NORTH	N ROCK CR, BLUE R, NW of Dillon	CR	SUM	1.0	11800	C 6	CUT
ROCK LAKE #2. NORTH	N ROCK CR, BLUE R, NW of Dillon	CR	SUM	2.5	11800	C 6	CUT
ROCK LAKE	LK FK CONEJOS R, SE of Platoro	RG	CON	5.0	9570	H 5	CUT
ROCK LAKE	ROCK CR, LEON CR, PLATEAU CR	CR	MES	3.0	10160	D 3	RBT
ROCK LAKE	ROCK CR, VALLECITO CR #2	SJ	HIN	30.9	11850	G 3	CUT
ROCK LAKE	TRIB TO ELK CR, CONEJOS R #2	RG	CON	7.2	9600	H 6	CUT
ROEBER POND	REYNOLDS CR, S of Paonia	GU	DEL	6.0	6960	E 3	CUT
ROELL LAKE	ROELL CR, VALLECITO CR #2	SJ	SNJ	4.6	12515	G 3	CUT
ROLLINS RESERVOIR	N FK WALLACE CR, S of Rulison	CR	MES	4.5	10070	D 3	CUT
ROSA LAKE	S FK MAD CR, NE of Mad Creek	YP	ROU	5.7	10000	A 5	CUT, BRK

Western Colorado Lakes

Name	Location	Drainage	County	Size	Elev.	Map Coor.	Fish Likely to be Present
ROUND LAKE	E FK WMS FK Y, SW of Yampa	YP	GAR	8.3	10350	B 4	CUT
ROWDY RESERVOIR	NE of Silverjack Res	GU	GUN	10.0	8000	F 3	RBT
ROXY ANNE LAKE, UPPER	RED CANYON, W of S Delaney Lk	NP	JAC	4.3	10400	A 5	CUT, RBT
ROXYANNE LAKE	RED CANYON, W of S Delaney Lk	NP	JAC	63.0	10204	A 5	CUT, RBT
RUBY LAKE, BIG	TEXAS CR, SW/Santa Maria Res	RG	MIN	29.6	11290	G 4	RBT, BRK
RUBY LAKE, LITTLE	TEXAS CR, SW/Santa Maria Res	RG	MIN	17.8	11250	G 4	RBT, BRK
RUBY LAKE, LOWER	RUBY CR, ANIMAS R, NE/Electra	SJ	SNJ	12.4	10820	G 3	CUT
RUEDI RESERVOIR	FRYING PAN R, 15 mi E/Basalt	CR	EAG	1000.0	7766	D 5	KOK, RBT, BRK, MAC, LOC
RUYBALID LAKE	ROUGH CR, CONEJOS R #2	RG	CON	6.9	11200	H 6	RBT, BRK
RYAN RESERVOIR	YOUNGS, WARD CR, S/Pine Res	GU	DEL	5.7	9940	D 3	RBT

S

Name	Location	Drainage	County	Size	Elev.	Map Coor.	Fish Likely to be Present
SABLE LAKE	N FK WHITE R, NW of Trappers L	WR	RBL	7.0	9882	B 4	CUT
SACKETT RESERVOIR	Grand Mesa, W/Wier&Johnson R	GU	DEL	9.4	10460	D 3	RBT, BRK
SALIDA HYDRO POND #2	S ARKANSAS R, NE/Monarch Pass	AR	CHA	1.0	8640	E 5	RBT, CUT
SAN CRISTOBAL LAKE	LK FK GUNNISON #3, S/Lake City	GU	HIN	346.0	8995	G 4	BRK, RBT, MAC
SAN FRANCISCO LK, UPPER W	SAN FRANCISCO CR, S/Del Norte	RG	RGR	4.2	11980	H 6	CUT
SANCHEZ LAKE, LOW	TRAIL CR, NE of Steamboat Lk	YP	ROU	4.6	10440	A 5	CUT, RBT
SANCHEZ LAKE, UPPER	TRAIL CR, NE of Steamboat Lk	YP	ROU	2.8	10640	A 5	CUT
SANDS LAKE	below Mt Shavano SFU	AR	CHA	6.0	6980	E 6	RBT, CUT
SAWTOOTH LAKE #2	TROUT CR, S of Sheriff Res	YP	RBL	2.0	10980	B 4	CUT
SAWTOOTH LAKE #3	TROUT CR, S of Sheriff Res	YP	RBL	3.0	10970	B 4	CUT
SAWTOOTH LAKE #4	TROUT CR, S of Sheriff Res	YP	RBL	3.0	10950	B 4	CUT
SAWTOOTH LAKE #6	TROUT CR, S of Sheriff Res	YP	RBL	2.0	10900	B 4	CUT
SELLAR LAKE	N FK FRYINGPAN R, SE of Biglow	CR	PIT	11.5	10220	D 5	RBT
SEVEN LAKES	S FK BIG CR, SW of Big Cr Lks	NP	JAC	14.0	10733	A 5	CUT
SEVEN SISTERS LAKE #3	HOMESTAKE CR, NW of Gold Park	CR	EAG	5.0	12150	D 6	CUT, BRK
SEVEN SISTERS LAKE #4	HOMESTAKE CR, NW of Gold Park	CR	EAG	5.0	12300	D 6	CUT, BRK
SEVEN SISTERS LAKE #5	HOMESTAKE CR, NW of Gold Park	CR	EAG	6.5	12750	D 6	CUT, BRK
SEYMOUR RESERVOIR	BUFFALO CR, SW of MacFarlane R	NP	JAC	32.8	8440	B 6	RBT
SHADOW LAKE	PATTERSON CR, S of Blair Lk	WR	GAR	6.0	10456	C 4	CUT, RBT
SHAFFER RESERVOIR	WILLOW CR, SW of Dunckley	YP	ROU	5.0	8830	B 4	RBT
SHALLOW LAKE	E MARVINE CR, W of Trappers Lk	WR	RBL	2.0	9560	C 4	BRK, CUT, RBT
SHAVANO POND	below Ridgway dam North pond	GU	OUR	1.8	6950	F 3	RBT
SHAW LAKE	LAKE CR, S FK RIO GRANDE R	RG	MIN	70.0	9860	G 3	CUT, RBT
SHEEP LAKE	S PRONG, CLIFF, COAL CR S	GU	GUN	8.9	10505	E 4	RBT, CUT
SHEEP SLOUGH RES	Grand Mesa	CR	MES	15.0	10060	D 3	RBT, CUT
SHEPHERD LAKE	SWEETWATER CR, S of Trappers L	CR	GAR	30.0	10762	C 4	CUT, MAC
SHERIFF RESERVOIR	TROUT CR, W of Yampa	YP	RBL	40.0	9723	B 4	RBT
SHERRY LAKE	LIME CR, NE of Biglow	CR	EAG	7.7	11000	D 5	BRK
SHINGLE LAKE	GYPSUM CR, EAGLE R	CR	EAG	2.1	10500	D 5	CUT, RBT, BRK
SHINGLE PEAK LAKE	S FK WHITE R, S of Trappers Lk	WR	GAR	6.0	11214	C 4	CUT
SHOESTRING LAKE	CHEDSEY CR, dwnstrm/Whale Lk	NP	JAC	7.2	10000	A 5	BRK
SIBERIA LAKE	N FK CRYSTAL R, NE of Marble	CR	GUN	3.3	11860	D 4	CUT
SILEX LAKE	STORMY GUL, VALLECITO CR #2	SJ	SNJ	6.8	12460	G 3	CUT
SILVER KING LAKE	PINE CR, W of Mt Harvard	AR	CHA	16.2	12640	E 6	CUT
SILVER LAKE	BIG CR, PLATEAU CR	CR	MES	19.0	10176	D 3	GRA, CUT
SILVER LAKE	BRIDAL VEIL CR, SW/Telluride	DO	SNM	4.0	11788	G 3	CUT
SILVERJACK RESERVOIR	CIMARRON R, S of Cimarron	GU	GUN	318.0	8900	F 3	RBT
SIOUX LAKE	LOSTMAN CR, N of Lost Man Res	CR	PIT	2.9	12080	D 5	CUT
SKILLET LAKE	BEAR R, SW of Yampa	YP	GAR	5.0	10710	B 4	BRK
SKINNY FISH LAKE	SKINNY FISH CR, N FK WHITE R	WR	GAR	58.8	10192	B 4	CUT, RBT, BRK
SLACK AND WEISS RES	COYOTE CR, SE of Seymour Lk	NP	JAC	14.8	8970	B 6	RBT
SLATE LAKE #3	SLATE CR. BLUE R #2	CR	SUM	9.6	11540	C 6	CUT, RBT, BRK
SLEEPYCAT PONDS	I5 mi E of Meeker	WR	RBL	3.0	6595	B 4	CUT
SLIDE LAKE #2	N FK ELK R HDWTRS	YP	ROU	2.1	10500	A 5	CUT
SLIDE LAKE #3	N FK ELK R HDWTRS	YP	ROU	4.8	10700	A 5	CUT
SLIDE LAKE #4	N FK ELK R HDWTRS	YP	ROU	4.2	11500	A 5	CUT
SLIDE LAKE	BIG BLUE CR HDWTRS	GU	HIN	5.2	10400	F 4	CUT
SLIDE LAKE	MARVINE CR, N FK WHITE R	WR	RBL	5.0	8650	C 4	CUT, RBT
SLIDE LAKE	N FK W TENNESSEE CR HDWTRS	AR	LAK	35.6	11725	D 6	CUT
SLIDE LAKE	W LAKE CR, EAGLE R, S/Edwards	CR	EAG	7.5	10700	C 5	CUT
SLIDE LAKE, LOWER	N FK TENN CR, NW Leadville	AR	LAK	10.0	10600	D 6	CUT, RBT
SLIDE LAKE, LOWER	NORRIS CR, Wet Big Rainbow Lk	NP	JAC	27.2	10527	A 5	CUT
SLOAN LAKE	LK FK GUNNISON R HDWTRS	GU	HIN	5.7	12920	G 3	CUT
SMITH LAKE, UPPER	BEAR R, N of Sweetwater Res	YP	GAR	15.0	10500	B 4	CUT, BRK
SNARE LAKE #1	SNARE CR, CTWNWD CR, W/Sherman	GU	HIN	1.6	12800	G 3	BRK
SNOW LAKE	PASS CR, S FK RIO GRANDE R	RG	MIN	1.2	10320	H 5	CUT
SNOWMASS LAKE	SNOWMASS CR, SW/Snowmass Village	CR	PIT	81.5	10980	D 4	RBT, CUT
SNOWSLIDE LAKE	BEAR R HDWTRS	YP	GAR	6.0	10790	B 4	RBT
SODA LAKE	SODA CR, N FK PINEY R, N/Vail	CR	EAG	2.0	10160	C 6	CUT
SOLITARY LAKE	MID FK DERBY CR, NW of Burns	CR	GAR	12.0	10638	C 5	BRK, RBT
SOUTH FORK RESORT POND	S FK WHITE R, SE of Buford	WR	RBL	2.0	7200	C 3	BRK
SOUTH MAMM PEAK LAKE	BEAVER CR, SE of Rulison	CR	GAR	2.0	10550	D 3	CUT
SPECTACLE LAKE	CONEJOS R #2, W of Antonito	RG	CON	4.2	8780	H 6	RBT
SPENCER LAKE	FRENCH CR, ALAMOSA R, E/Platoro	RG	CON	11.4	9820	H 6	RBT

Name	Location	Drainage	County	Size	Elev.	Map Coor.	Fish Likely to be Present
SPLAINS GULCH LAKE	SPLAINS GUL, SW/Crested Butte	GU	GUN	7.7	10381	E 4	CUT
SPRING CREEK POND	Spring Creek Pass, SE/Brown Lk	RG	HIN	7.9	9200	G 4	RBT
SPRING CREEK RESERVOIR	SPRING CR, NE of Almont	GU	GUN	86.9	9915	E 5	RBT, CUT, BRK
SPRING LAKE	SAND CR, TROUT CR	YP	RBL	3.0	10080	B 4	RBT
SPRUCE LAKE, LOWER	S FK RG R, SW/Big Meadows Res	RG	MIN	19.8	11100	G 5	CUT
SPRUCE LAKE, UPPER	S FK RG R, SW/Big Meadows Res	RG	MIN	19.8	11120	G 5	CUT
ST KEVIN LAKE	BEAR CR, N of Turquoise Lk	AR	LAK	4.2	11850	D 6	CUT
STAGECOACH RESERVOIR	YAMPA R #6, E of Oak Creek, State P	YP	ROU	700.0	7500	B 5	RBT, CUT, KOK SPL, WHF
STAMBAUGH RESERVOIR	CROSBY CR, SW of Coalmont	NP	JAC	10.0	8880	A 5	RBT
STEAMBOAT LAKE	WILLOW CR, NW of Steamboat Sprg	YP	ROU	1081.0	8000	A 5	RBT, CUT, LOC
STILLWATER RESERVOIR	BEAR R, SW of Yampa	YP	GAR	88.0	10255	B 4	RBT, BRK, CUT
STRAWBERRY LAKE #1	LIME CR HDWTRS, NE of Biglow	CR	EAG	5.0	11220	D 5	BRK
STRAWBERRY LAKE #2	LIME CR HDWTRS, NE of Biglow	CR	EAG	10.0	11260	D 5	BRK
SUGARLOAF LAKE	GYPSUM CR, EAGLE R, W/Rim Lk	CR	EAG	4.0	10700	A 5	RBT, BRK, CUT
SUMMIT RESERVOIR	NE of Corlez	DO	MTZ	351.0	7368	H 2	RBT, CCF, NPK, WAL, CRA
SUNLIGHT LAKE, LOWER	SUNLIGHT VALLECITO CR #2	SJ	LAP	17.0	12033	G 3	CUT
SUNLIGHT LAKE, UPPER	SUNLIGHT VALLECITO CR #2	SJ	LAP	14.4	12545	G 3	CUT
SUNNYSIDE LAKE #1	SUNNYSIDE CR, NW of Burns	CR	ROU	5.0	10360	C 4	BRK
SUNNYSIDE LAKE #2	SUNNYSIDE CR, NW of Bums	CR	ROU	7.0	10360	C 4	BRK
SUNNYSIDE LAKE #3	SUNNYSIDE CR, NW of Burns	CR	ROU	2.5	10380	C 4	BRK
SUNSET LAKE	Mesa Lakes Resort, Grand Mesa	CR	MES	18.0	9790	D 2	RBT, BRK, LOC
SUPPLY BASIN RES #1	DEEP CR, NW of Dotsero	CR	GAR	18.5	10780	C 4	RBT
SURPRISE LAKE	FRASER CR, S E of Trappers Lk	WR	GAR	9.0	11128	C 4	CUT, RBT
SWALE LAKE	PARK CR, S FK RIO GRANDE R	RG	RGR	5.4	9895	H 5	RBT
SWEDE LAKE	S FK WHITE R, SE of Buford	WR	RBL	5.0	8880	C 3	BRK, RBT
SWEETWATER LAKE	25 mi NW of Dotsero	CR	GAR	72.0	7709	C 4	RBT, KOK
SYLVAN LAKE	15 miles SE of Eagle, State Park	CR	EAG	42.0	8510	D 5	RBT, BRK

T

Name	Location	Drainage	County	Size	Elev.	Map Coor.	Fish Likely to be Present
TABLE LAKE	LECHE CR, RIO BLANCO	SJ	ARC	4.3	9600	H 5	RBT
TABOR LAKE	LINCOLN CR, SE of Aspen	CR	PIT	5.7	12320	D 5	CUT
TAYLOR LAKE	HDWTRS TAYLOR R	GU	GUN	19.3	11544	D 5	BRK, RBT
TAYLOR RESERVOIR	Taylor Park, NE of Almont	GU	GUN	2009.0	9330	E 5	RBT, CUT, KOK, MAC
TEAL LAKE	BEAVER CR, W of Coalmont	NP	JAC	15.6	8812	A 5	RBT
TELLURIUM LAKE	LAST CHANCE CR, SE of Woods Lk	CR	EAG	7.5	10535	D 5	CUT
TEN MILE LAKE	TEN MILE CR, ANIMAS R #3	SJ	SNJ	7.8	12264	G 3	CUT
TENNESSEE LK, W, UPPER	W TENN CR, NW of Leadville	AR	LAK	18.8	11800	D 6	CUT, BRK
TENNESSEE LK. W, LOWER	W TENN CR, NW of Leadville	AR	LAK	4.0	11775	D 6	CUT, BRK
TERRELLS LAKE	LOST MAN CR, ROARING FK R	CR	PIT	3.6	12390	D 5	CUT
TERRIBLE CREEK POND	ROUBIDEAU CR, SE/Columbine Pass	GU	MON	3.0	9200	F 2	BRK
TEXAS LAKE #2	TEXAS CR, E of Taylor Res	GU	GUN	5.9	9827	E 6	RBT, BRK
TEXAS LAKE #3	TEXAS CR, E of Taylor Res	GU	GUN	6.0	9827	E 6	RBT, RBT
TEXAS LAKE, NORTH	N TEXAS CR, TAYLOR R #2	GU	GUN	5.6	12050	E 6	CUT
THOMAS LAKE #1	PRINCE CR, CRYSTAL R, SW/Basalt	CR	PIT	9.2	10200	D 4	CUT
THOMAS LAKE	E FK LAKE CR, EAGLE R. S/Edward	CR	EAG	14.0	12675	C 5	CUT, RBT
THOMPSON LAKE	L FK GUNNISON R, NW/Lake City	GU	HIN	4.3	9920	F 4	CUT, BRK, RBT
THREE LICKS LAKE	THREE LICKS CR, SHEEPHORN CR	CR	EAG	2.0	10030	C 3	CUT
TIAGO LAKE	BEAVER CR, W of Coalmont	NP	JAC	8.6	8860	A 5	RBT
TIMBER LAKE	S FK CONEJOS R, SW of Platoro	RG	CON	11.9	11322	H 5	CUT, RBT
TIPPERARY LAKE, LOWER	CATARACT CR, SW/Green Mtn Res	CR	SUM	5.0	9760	C 6	RBT
TOBACCO /TOBASCO) LAKE	SADDLE CR, CONEJOS R. SW/Platoro	RG	CON	12.8	12280	H 5	RBT, CUT
TOMAHAWK LAKE	DUGOUT CR, N of Paonia Res	GU	DEL	16.0	8300	D 4	BRK
TOTTEN RESERVOIR	E of Cortez	DO	MTZ	204.0	6158	H 1	CCF, LMB, NPK, WAL
TRAIL LAKE	CANON ESCONDIDO, S FK CONEJOS	RG	CON	29.6	11987	H 5	RBT
TRAIL LAKE	MARVINE CR, N FK WHITE R	WR	RBL	3.0	10500	C 4	RBT
TRAPPERS LAKE, LITTLE	CABIN CR, N FK WHITE R	WR	GAR	20.0	9926	C 4	CUT
TREASURE VAULT LAKE	CROSS CR HDWTRS, EAGLE R	CR	EAG	10.0	11675	D 5	CUT
TRICKLE PARK RESERVOIR	Grand Mesa, E/Youngs Cr Res	GU	DEL	123.0	9928	D 3	RBT
TRINITY PEAKS LAKE	TRINITY CR, VALLECITO CR #2	SJ	SNJ	7.0	12396	G 3	RBT
TRIO RESERVOIR	E FK SURFACE, E/Cedar Mesa Res	GU	DEL	17.0	10200	D 3	CUT
TROUT LAKE	Lizard Head Pass, SW/Telluride	DO	SNM	126.5	9714	G 3	RBT, BRK, CUT
TROUT LAKE	W TROUT CR HDWTRS, SW of Creede	RG	HIN	23.7	11685	G 4	CUT
TROUT MEADOWS POND	James Ranch, E of Hermosa	SJ	LAP	2.0	6600	H 3	RBT
TROUT RESERVOIR	Grand Mesa, S of Elk Park Res	GU	DEL	8.4	10280	D 3	RBT
TRUJILLO MEADOWS RES	RIO DE LOS PINOS, N of Cumbres	RG	CON	69.2	10020	H 6	RBT, LOC, CUT
TRURO LAKE	LINCOLN CR, SE of Aspen	CR	PIT	6.7	12190	D 5	CUT
TUCKER PARK PONDS	NE of Wolf Creek Pass	RG	MIN	9.9	9640	H 5	RBT
TUHARE LAKE, LOWER	FALL CR. EAGLE R, SW/Red Cliff	CR	EAG	12.0	12090	D 6	CUT
TUHARE LAKE. UPPER	FALL CR. EAGLE R, SW/Red Cliff	CR	EAG	43.0	12365	D 6	CUT
TUNNEL LAKE	TUNNEL GUL, CHALK CR, SW/St EIm	AR	CHA	6.5	11952	E 6	CUT
TURQUOISE LAKE	W of Leadville	AR	LAK	1650.0	9869	D 5	RBT, LOC, CUT, MAC
TWELVEMILE LK, LOWER	TWELVEMILE CR, SW of Fairplay	SP	PAR	8.0	11097	D 6	CUT, BRK
TWELVEMILE LK, UPPER	TWELVEMILE CR, SW of Fairplay	SP	PAR	3.8	11940	D 6	CUT, BRK
TWILIGHT LAKE, LOWER	LIME CR, CASCADE CR, E/Potato	SJ	SNJ	4.2	11570	G 3	RBT, BRK
TWILIGHT LAKE, NORTH	LIME CR, CASCADE CR, E/Potato	SJ	SNJ	2.8	12050	G 3	RBT, BRK
TWILIGHT LAKE. SOUTH	LIME CR, CASCADE CR, E/Potato	SJ	SNJ	2.5	11700	G 3	RBT, BRK

Name	Location	Drainage	County	Size	Elev.	Map Coor.	Fish Likely to be Present
TWIN LAKE #1, NORTH	Grand Mesa, N of Bonita Res	GU	DEL	23.2	10400	D 3	RBT
TWIN LAKE #2 SOUTH	Grand Mesa, N of Bonita Res	GU	DEL	20.5	11520	D 3	RBT
TWIN LAKE, LOWER #2	MID BRUSH CR, NE/Crested Butte	GU	GUN	3.2	11780	E 5	RBT, CUT
TWIN LAKE, SOUTH	PINE CR, W of Mt Harvard	AR	CHA	5.6	12250	E 6	CUT
TWIN LAKE, UPPER #1	MID BRUSH CR, NE/Crested Butte	GU	GUN	8.1	11790	E 5	RBT, CUT
TWIN LAKE, UPPER (W)	CANON RINCON, S FK CONEJOS	RG	CON	2.0	11740	H 5	RBT
TWIN LAKES	SW of Leadville	AR	LAK	1.0	9200	D 6	RBT, CUT, MAC
TWIN LAKES, LOWER	SW of Leadville	AR	LAK	2000.0	9200	D 6	RBT, CUT, MAC
TWIN MEADOWS POND	FOSTER GULCH, FRYING PAN R	CR	PIT	10.0	9600	D 5	RBT
TWO LEDGE RESERVOIR	NINEGAR CR, SE of Seymour Lk	NP	JAC	5.7	9740	B 6	RBT
U							
UTE LAKE, LOWER TWIN(#2)	MID UTE CR, SW of RG Res	RG	HIN	4.7	11792	G 4	CUT
UTE LAKE, MAIN(E)	UTE CR, SW of RG Res	RG	HIN	32.1	11847	G 4	CUT
UTE LAKE, MIDDLE	MID UTE CR, SW of RG Res	RG	HIN	11.4	11949	G 4	CUT
UTE LAKE, UPPER (W)	W UTE CR, SW of RG Res	RG	HIN	4.7	12326	G 4	CUT
UTE LAKE, UPPER TWIN(#1)	MID UTE CR, SW of RG Res	RG	HIN	15.8	11792	G 4	CUT
V							
VALLECITO LAKE	VALLECITO CR #2 HDWTRS	SJ	SNJ	16.0	12010	G 3	CUT
VALLECITO RESERVOIR	LOS PINOS R, N of Bayfield	SJ	LAP	2718.0	7664	H 3	RBT, CUT, NPK, LOC
VAUGHN LAKE	POOSE CR, N of Trappers Lk	YP	RBL	36.0	9390	B 4	RBT
VEGA RESERVOIR	PLATEAU CR, E of Collbran, State P	CR	MES	900.0	7985	D 3	RBT, CUT, LOC, BRK
VELA RESERVOIR	Grand Mesa, N/Trickle Park Res	GU	DEL	18.8	10160	D 3	RBT
VERDE LAKE LITTLE/LOW	ELK CR, ANIMAS R #3	SJ	SNJ	3.2	12160	G 3	RBT, CUT, BRK
VERDE LAKE, BIG/UPPER	ELK CR, ANIMAS R #3	SJ	SNJ	9.8	12165	G 3	RBT, CUT, BRK
VESTAL LAKE	ELK CR, ANIMAS R #3	SJ	SNJ	3.9	12260	G 3	RBT, CUT, BRK
W							
WALL LAKE	N FK WHITE R, above Trappers L	WR	GAR	40.0	10986	C 4	CUT
WALTERS LAKE	LONE LICK CR, SHEEPHORN CR	CR	EAG	8.0	10360	C 6	RBT, BRK, CUT
WARD CREEK RESERVOIR	Grand Mesa	GU	DEL	26.0	9800	D 3	RBT, CUT
WARD LAKE	WARD CR, E of Island Lk	GU	DEL	84.0	10110	D 3	RBT, BRK, SPL
WATER DOG RESERVOIR'	COON CR, Grand Mesa	CR	MES	24.0	9930	D 2	RBT
WATERDOG LAKE #1	W GROUSE CR. EAGLE R, SW/Minturn	CR	EAG	2.0	10760	C 5	CUT
WATERDOG LAKE	LK FK GUNNISON R, E/Lake City	GU	HIN	6.9	11120	F 4	BRK
WATERDOG LAKE	S ARKANSAS R, W of Monarch	AR	CHA	15.0	11475	E 6	CUT, BRK
WATTENBERG PONDS	BROWN CR, W of Lake John	NP	JAC	4.0	9360	A 5	RBT
WEBB LAKE	NEEDLE CR, NE of Electra Lk	SJ	LAP	6.3	10942	G 3	CUT, RBT
WEIR & JOHNSON RES.	Grand Mesa, NE of Bonita Res	GU	DEL	45.9	10480	D 3	RBT, BRK
WELLER LAKE	ROARING FK R, SE of Aspen	CR	PIT	8.6	9550	D 5	RBT
WEST FORK LAKES (2)	W FK SAN JUAN R	SJ	MIN	2.0	8000	H 5	RBT
WESTON PASS LAKE	S FK of SP R, SE of Weston Pass	SP	PAR	1.5	11840	D 6	CUT
WHEAT LAKE	TROUT CR, S of Sheriff Res	YP	RBL	2.0	9930	D 4	CUT
WHEELER LAKE #1	OFFICERS GULCH, SW/Dillon Res	CR	SUM	6.0	11050	C 6	CUT
WHEELER LAKE, LOWER	PLATTE GUL, MID FK SP, NW/Alma	SP	PAR	28.0	12183	D 6	CUT
WHEELER LAKE. UPPER	PLATTE GUL, MID FK SP, NW/Alma	SP	PAR	4.0	12450	D 6	CUT
WHITE DOME LAKE	ELK CR, ANIMAS R #3	SJ	SNJ	5.5	12560	G 3	RBT
WHITNEY LAKE #1	HOMESTAKE CR, EAGLE. N/Gold Pr	CR	EAG	5.3	10956	D 6	BRK, RBT, CUT
WILLIAMS CREEK RESERVOIR	WILLIAMS CR, above Paonia Res	GU	GUN	12.0	8180	E 4	RBT
WILLIAMS CREEK RESERVOIR	WILLIAMS CR, N of Pagosa Spgs	SJ	HIN	343.0	8241	G 4	KOK, CUT, RBT, BRK
WILLIAMS FORK RESERVOIR	WMS FK COLO R, SW of Parshall	CR	GRA	1600.0	7811	B 6	RBT, KOK, NPK
WILLIAMS LAKE	E SOPRIS CR, W/Snowmass Village	CR	PIT	11.6	10815	D 4	BRK, CUT
WILLIAMS LAKE	WILLIAMS CR HDWTRS, PIEDRA R	SJ	HIN	11.5	11695	G 4	CUT
WILLIS LAKE	WILLIS GUL, LAKE CR, SW/Twin L	AR	CHA	22.3	11800	D 6	CUT
WILLOW LAKE	WILLOW CR, Maroon Cr. SW/Aspen	CR	PIT	19.3	11795	D 5	BRK, CUT
WILMOR POND	I-70 W of Edwards	CR	EAG	2.0	7105	C 5	RBT
WINDSOR LAKE	BUSK CR, SW of Turquoise Lk	AR	LAK	43.0	11625	D 6	CUT
WINDY POINT LAKE	MARVINE CR, N FK WHITE R	WR	RBL	5.5	10700	C 4	CUT
WOLFORD MTN. RES	MUDDY CREEK, N of Kremmling	CR	GRA	1400.0	7550	B 6	RBT
WOLVERINE LAKE	WOLVERINE CR, S FK ELK R	YP	ROU	7.3	10284	A 5	RBT, CUT
WOODS LAKE	DOME CR, BEAR R	YP	GAR	5.0	9700	B 4	CUT
WOODS LAKE	MUDDY CR, S of Sawpit	DO	SNM	19.8	9423	G 2	RBT, BRK
WRIGHTS LAKE	CHALK CR, SW of Nathrop	AR	CHA	2.0	8380	E 6	RBT
WRIGHTS LAKE	SNEFFLES CR, CANYON CR, UNC R	GU	OUR	1.5	12200	G 3	CUT
Y							
YAMCOLO RESERVOIR	BEAR R, SW of Yampa	YP	GAR	175.0	9680	B 4	RBT, LOC, BRK, WHF
YELLOW LAKE	E FK NO NAME CR, N/Glenwood Sp	CR	GAR	8.0	10380	C 4	BRK
YOUNGS CREEK RES #1 & 2	Grand Mesa, SE/Big Eggleston Res	GU	DEL	52.1	10180	D 3	RBT, BRK
YOUNGS CREEK RES #3	Grand Mesa, E/Big Eggleston Res	GU	DEL	23.0	10200	D 3	RBT
YOUNGS LAKE	Grand Mesa	GU	DEL	10.0	10300	D 3	BRK
YULE LAKE #1	YULE CR, CRYSTAL R, SE/Marble	CR	GUN	7.0	12140	D 4	CUT
YULE LAKE #2	YULE CR, CRYSTAL R, SE/Marble	CR	GUN	7.6	11910	D 4	CUT
YULE LAKE #3	YULE CR, CRYSTAL R, SE/Marble	CR	GUN	1.4	11920	D 4	CUT
YULE LAKE #4	YULE CR, CRYSTAL R, SE/Marble	CR	GUN	3.3	11840	D 4	CUT
YULE LAKE #5	YULE CR, CRYSTAL R, SE/Marble	CR	GUN	10.0	11860	D 4	CUT
Z							
ZEIGLER RES	GRASSY CR, SE of Hayden	YP	ROU	10.0	6735	B 4	RBT

Index of Lakes and Reservoirs

Index of Lakes and Reservoirs

Index of Lakes and Reservoirs